Y0-BRR-808

STATISTICS
AT THE SCHOOL LEVEL

STATISTICS AT THE SCHOOL LEVEL

The Interpretation of Official Statistics

Proceedings of the Third International Statistical
Institute Round Table Conference
on the Teaching of Statistics

Edited by
[*Lennart Råde*]
Chalmers University of Technology
Gothenburg – Sweden

ALMQVIST & WIKSELL INTERNATIONAL
Stockholm – Sweden

———

A Halsted Press Book
John Wiley & Sons
New York – London – Sydney – Toronto

© 1975 International Statistical Institute
Voorburg, The Netherlands

First published in Sweden by
Almqvist & Wiksell International AB, Stockholm

ISBN 91 – 2200004 6

Published in the Western Hemisphere and the United Kingdom by
Halsted Press, a Division of John Wiley & Sons, Inc., New York

Library Congress Catalog Card No.:
Halsted Press ISBN 0 470 –

Printed in Sweden by
Göteborgs Offsettryckeri, Stockholm 1975

Contents

Preface

In the framework of the International Statistical Education Programme which is undertaken by the International Statistical Institute (ISI), under the auspices of and with support from UNESCO, a series of Round Table Meetings on the teaching of statistics has been organized by ISI. As a satellite meeting associated with the 1973 biennal ISI meeting, the third such round table conference was held in Vienna from August 30 to September 4 1973. The theme of the conference was *the teaching of statistics at the secondary school level.*

These proceedings contain the recommendations of the conference and the papers presented at the meeting. The first of the papers (by Benjamin, Goldberg, and Keyfitz) deals with the teaching of official statistics. The papers by Engel, Rao, and Whitfield deal with new methods and attitudes and the remaining papers (by Halmos, Hennequin, Kruskal, Lang, Råde, Rucker, Oyelese, and Postelnicu) give reports on activities and educational experiments in the teaching of statistics from many different parts of the world.

It is a pleasure for me to thank Frederick Mosteller on behalf of the participants for his leadership of the conference, executed with great charm and effectiveness. Our thanks are also due to E. Lunenberg, director of ISI, and Agnes Whitfield, secretary of the conference, for their careful planning of the conference. The local arrangements were made in a most pleasant way by the Stadtschulrat für Wien under the responsibility of dr Leopold Peczar. Finally it is a great pleasure for me to thank Elisabeth Klenner of Vienna for her expert help in the preparation of this volume.

Göteborg, Sweden, May 1974

Lennart Råde

RECOMMENDATIONS OF THE CONFERENCE

As a result of the papers presented at the 3rd Round
Table Conference and of the discussions among the members,
we offer a variety of recommendations. These recommenda -
tions fall into four parts:

I.   Creation of new materials,
II. Teacher training,
III. Conferences, and
IV. Other suggestions dealing with various
     attitudes, actions, and surveys.

The Round Table has generally assigned priorities according to the order given within each part, but has not
assigned relative priorities across parts.

We believe these tasks can profitably be carried out by a
variety of agencies. The task in some recommendations
could well be executed by a single individual or by a
small group. Some could be carried out by a national group
and still others would require the effort of several na -
tional groups or even the effort of an international or -
ganization. For some recommendations the Round Table naturally looks to the Committee on Education of the International Statistical Institute as the permanent agency likely
to be able to follow such tasks through.

I. CREATION OF NEW MATERIALS

1. Official statistics.

Governmental or official statistics inevitably control
portions of the lives of every citizen of every country,
and our ability or inability to change these statistics
through public policy actions set goals and limitations
for each society. Consequently citizens should be in -
formed of the vital nature of these quantities and of
their interdependencies. In the past these important

ideas have been largely the concern of specialists ,
but we believe that now a wider understanding of these
ideas could benefit every society, recognizing , of
course, that different countries have differing needs
and that people of varying ages and interests may need
differing forms of instruction.

As an example of the kind of material that a task force
might prepare, we suggest discussions of the statistical
information requirements needed to plan, execute, and
monitor effects of changes in public policy designed to
solve a particular problem. The area would be chosen by
the task force, but examples might be "tuning the economy",
changes in educational policy to achieve some specific end,
or steps in the delivery of health care required, say, to
increase longevity or reduce neo-natal mortality. Such
concrete problems require a variety of kinds of information. The reader would learn a number of important sta -
tistical ideas (though we do not intend that they get in
this way a systematic course in quantitative methods) ,
and they would learn how some things in their own lives
relate to these official statistics. Naturally, these materials must be prepared in an inviting and challenging
manner. We anticipate a sequence of pamphlets of modest
size published simultaneously in several languages. Although
the general plan is intended for the use of secondary schools,
this covers a considerable age range and the material pre -
pared for younger students need not dig so deeply and rea -
listically into the problems as that prepared for the older
groups.

2. Materials for school administrators and those in similar positions

We recommend the preparation of a short, friendly, informative pamphlet on the use of statistics in a variety of
fields. In conjunction with this, a panel of speakers would
be formed and made available for presenting, at national
assemblies of administrators, talks at major sessions on

the uses of statistics throughout the world. (Some illustrative materials of this kind are contained in Statistics: A Guide to the Unknown though the govern - mental examples are keyed primarily to the United States of America. These would need to be supplemented for the broader audience intended here.)

3. Material for students

We recommend:

    a) Preparation of a brochure describing the utilitarian value of statistics in scientific and non-scientific disciplines and in everyday understanding of current events.

    b) Preparations of series of tasks which a student should be able to accomplish at the conclusion of a course in elementary statistics at the secondary school level.

4.Lesson sketches

It is recommended that a group of 10 to 20 statisticians and teachers who have experience in teaching and writing meet to produce in one to two weeks a book with the title Some Lessons in Statistics. The book of three or four hundred pages should be published as paper-back. It should contain sketches of lessons in sufficient detail that a teacher can teach them almost directly from the sketch. There should be lessons for all secondary school grades and different levels of difficulty. The multiple author- ship will automatically guarantee variety in content and level. There are excellent examples of books of this type (produced in the way outlined above), for instance:

    T. J. Fletcher, Ed., Some Lessons in Mathematics, A Handbook on the Teaching of "Modern" Mathematics, Cambridge University Press, 1964

    D. H. Wheeler, Ed., Notes on Mathematics in Primary Schools. Association of Teachers of Mathematics, 1967

5. Additional real-life examples

The collection of four volumes of real-life statistical examples together with exercise material and teacher's manuals, entitled Statistics by Example, is prototypic of a kind of material that the Round Table wishes to have more widely available and better codified.

In order to be internationally useful, these materials should have the following characteristics:

1) The examples should be categorized by: a) disciplines and b) applications within each dis - cipline.

2) In those instances where gaps are detected within either a discipline or in applied areas, appro - priate materials should be supplied.

3) Especially (but not exclusively) for developing countries, translations should be made and the document would then be printed inexpensively for distribution to appropriate schools and/or other interested groups.

In preparing materials, the nature of activities relevant to the various countries should be taken into account. Example: For any item dealing with equipment and requiring electricity for operation, attention must be given to the fact that in many schools in some countries electricity is not available.

Who will produce the material to fill the gap? First of all it should be noted that items, especially of the types under consideration, are not readily available. This is not to infer that such items cannot be obtained. The Round Table recommends that a group of individuals selected from several countries be asked, well ahead of time, to write up items of a specific nature. These individuals would submit their items to an individual who would, in turn, be responsible for the selection and editing of needed items. There may be value in calling a conference to assist in editing these materials so that they can be useful in more countries. Minor revisions may make items much more widely available by avoiding culture-bound expositions.

6. Audio-visual aids.

We recommend that consideration be given to the development of interdisciplinary statistical learning packages, in a number of languages, which would be used with audio-visual aids.

This method of teaching would: a) encourage group work , b) demonstrate in an attractive and exciting manner statistical concepts and ideas, c) be relatively inexpensive as the visual material is nearly universal and only dubbing of languages and translation for the accompanying text would require adaptation to national requirements.

7. Experimental investigations.

The interface between statistics and other disciplines is extremely important for progress of research both in statistics and other disciplines. Teaching of statistics without explaining its relevance to practical problems may end only in solving mathematical exercises. Concepts of randomness, levels and errors of measurements, population and sample, estimation of unknown quantities and relationships, testing of hypotheses, etc., which are in - volved in scientific inference and decision making, can well be explained and demonstrated through careful choice of practical problems arising in some branches of science and providing an opportunity for students to collect live data.

The contents of the course in statistics at the high school level and the method of presentation should be designed more to expose the students to quantitative thinking , inference under uncertainty, and to inculcate in them a spirit of inquiry. The late professor J. B. S. Haldane laid great emphasis on an interdisciplinary approach to

the teaching of statistics and suggested some experiments
involving measurements and collection of data for explaining
the theory and methods of statistics.

It would be valuable to make a list of experimental investigations in different branches of science --investigations
that can be easily carried out with students' participation
and that provide live data for discussing and illustrating
statistical methods. An investigation of diurnal variation
in an individual's stature presented at the Round Table
would be a suitable example (see paper by C. R. Rao). A
manual may be prepared describing how these investigations
can be planned, what data can be collected, what methods
can be used for analysis of data and what aspects of statistics can be developed around each experiment. The manual
which might be titled Statistics, an Interdisciplinary
Approach, would be a good supplement to Statistics by Example.

8. Field trips.

In many countries it is customary to take school classes
on expeditions or field trips to museums, concerts, factories,
other parts of the nation, and the like. An informal survey
of six nations represented at the Round Table found a great
variety of field trips in existence, trips with different
objectives, organization, and frequency. In Nigeria, for
example, such trips are typically organized by school clubs,
e.g., the geography club. In Austria and Germany, some trips
stress physical education--e.g., hiking or skiing--and others
are academic. Trips may range in lenght from a few hours
to two weeks. There was a general view that field trips
contribute considerably to socialization of school groups.
Among the trips reported by the participants few dealt
with the mathematical sciences and hardly any had a statistical character.
Several participants emphasized the severe requirements
in organizing a useful statistical field trip: careful
preparation, a knowledgable local informant, follow-up
discussions, whenever possible relevant activity by the

students (e.g., trying out actual simple physical inspection as in a factory quality control program).

Against this background we recommend the following ISI activities:

a. Survey. A more extensive survey of field trips might be carried out. There should be special reference to statistics and probability, and the survey should cover field trips actually carried out. The survey should also gather information about field trips not dealing with statistics and probability but onto which could readily be added statistical and probabilistic components. For example, in Germany, a class may go to a country village for two weeks and study its economic, agricultural, and geographical circumstances. It would be straightforward to introduce basic statistics in such a study.

b. Publication. Subsequent to completion of 8a, immediately above, we recommend formation of a writing team to put together descriptions of successful field trips in a variety of countries and cultures.

Finally, we note that there is no sharp line between the idea of field trips and that of projects, as recommended elsewhere in this report. Indeed, a field trip is likely to be enhanced by combining it with a project, and vice versa. For example, a visit to a factory's quality control unit might be combined with a quality control project in a school's shop or cafeteria.

II. TEACHER TRAINING

1. International education program.

Introduction of statistics at the secondary level will not yield desired results unless teaching is done by qualified teachers well versed in theoretical statistics and with some experience in practical applications. There is bound to be a shortage of such teachers for years to come.

We suggest that the ISI initiate and organize programs for training teachers at the international level. This is to encourage a high standard of education in statistics in different countries while recognizing the varying needs of different countries. (See Recommendation 2, immediately below.)

What sort of facilities are needed to accommodate an international institute such as described above? It was learned, directly subsequent to the discussion of this recommendation that Austria, for example, sometimes has such a facility available, including lecture halls, small group rooms, living quarters, library facilities and recreational areas, to accomodate 40- 50 participants.

2. For training of teachers (national).

a) Pre-service: Preparation of a syllabus (and rationale therefore) for a course or courses at the undergraduate level which would serve the needs of those who may teach statistics at the secondary level. Such a course would be suggested as a requirement for fulfilling the degree specifications for prospective teachers of: chemistry, physics, biology, earth sciences, mathematics, and all of the social sciences.

b) Inservice: Preparation of an outline for statistics which when studied as a course would give teachers, already in service but without statistical training, the necessary background to teach an elementary course in statistics at the secondary level.

III. FUTURE ROUND TABLE CONFERENCES

We recommend that the sequence of ISI Round Tables on Teaching Statistics be continued. As possible themes for future Round Tables we suggest four topics. The Round Table regards the first two items as of equal importance and generally of higher urgency than the last two which are also regarded as equally important.

1. Teaching of probability and statistics at the primary school level.

2. Joint Round Tables with other international organizations (physics, biology, medicine, mathematics, etc.).

3. Teaching of statistics to adults:

      a) general culture,
      b) younger adults who find a need to know
         statistics late in their education
         (physicians, administrators, etc.).

4. Consideration of problems of teacher-training in order to achieve some of the objectives implied by the three topics 1,2, and 3 listed immediately above.

IV. ATTITUDES, ACTIONS, AND SURVEYS

1. Cooperation between professional statisticians and training groups.

It is important that statisticians and statistical organizations be concerned with the teaching of statistics at the secondary school level. It is therefore recommended that, in committees working with such teaching programs, pro - fessional statisticians be involved and that contacts and cooperative efforts be established with statistical or - ganizations.

2. Experimentation in statistical education.

We recommend that an adequate portion, perhaps 5%, of the resources devoted to trials of new teaching methods be used for evaluative studies of those methods. In particular, the evaluations should, wherever feasible, be based on proper experiments, with randomization and control groups.

Background: careful evaluation is important in order to provide clear feedback and a cumulatively increasing body of knowledge. Without substantial resources spent on evaluation and dissemination of its results, it is all too easy

to find the same apparently novel approach tried time
after time, in place after place, without knowledge of
prior or separate trials.

Broadly speaking, there are two kinds of evaluative investigations. First, there is the descriptive survey; this
includes a careful statement of the activities under study,
perhaps interviews with teachers and students, perhaps
some follow up both of individuals in the study and of
numbers enrolled in classes and the like.

Second, there is the proper experiment, with its randomized
assignment to control and trial groups. Both kinds of study,
when conscientiously done, face common difficulties. For
example, there is the so-called Hawthorne effect: people
often behave differently than otherwise when they know themselves to be participants in a special study. It is always
difficult to decide on criteria and on the relevant time
scale: scores on what examination at the end of the course?
one year later?

A proper experiment has additional difficulties. One must
decide on the basic units (students?, classes?, school?,
cities?). There may be various levels of control. Maintaining effective independence among units is often trouble-
some. Yet it is the proper experiment that permits con -
clusions of known confidence and allows us to gain sharp
informations on causative relations.

Large-scale experiments for new social and pedagogical
programs may be very expensive, and we do not suggest de-
laying promising trials until resources can be found for
proper experimental evaluations. Yet there are many small
or intermediate scale programs for which reasonable experiments could be practicably incorporated. For example,
there have been several curricular suggestions-- even
completed textbooks-- whose aim is to motivate and awaken
interest in students who have strongly negative views
about mathematics and have correspondingly weak skills. The
essential notion is to work on the observation and nume -
rical side of statistical questions having direct interest
to such students. Since groups of mathematically blank

students may be found in profusion, middled-sized experiments with controls and randomization might be quite practicable.

3. Computers.

Computers or even small inexpensive electronic calculators would extend significantly the range of problems that can be treated. We recommend that efforts should be made to provide schools in each country with computers and/or calculators so that innovations could begin as soon as possible.

4. International survey.

To determine how extensively statistics is currently taught in secondary schools throughout the world, a survey should be made. Such a survey can be accomplished by means of an instrument, carefully worded so that each item would be universally understood. The instrument should be sent to the following school personnel: teachers, those responsible for curriculum development, and certain policy-making groups. The result of the survey should reveal information about:
   a) the teacher,
   b) the student,
   c) the course,
   d) the facilities available for experimental activities
      by the students,
   e) some outstanding projects in statistics,
   f) the availability in colleges and universities of
      courses appropriate for prospective teachers of
      statistics.
Such an instrument prepared especially for the United States may be found in the paper presented by I. Rucker to this conference.

A single organization could take on the task of disseminating the survey, receiving responses to it, and analyzing and reporting the results. It is recommended that the Committee on Education of the ISI take steps to have the survey executed. We note the need to involve people who

work at the secondary level. The information, when written
up, would be sent along with pertinent recommendations of
the 3rd Round Table Conference on the Teaching of Statistics to educational authorities, mathematical and statistical associations, and other professional bodies in
different countries.

5. <u>International communication of teaching experience.</u>

It is essential that there should be media for the regular
international communication of attitudes to teaching ,
experience with and results of new methods, guidelines to
future developments, as well as suggestions about school
syllabuses and textbooks or other teaching material and
aids. We recommend that the ISI should accept responsibility
for maintaining this communication which could take different forms; for example, there could be some form of news
letter which could be circulated not only to ISI members
but also to educational institutions and teachers organizations; there could be a regular provision of space in the
<u>ISI Review</u> (helping thereby to expand its circulation) ;
or space could be sought by the ISI in other journals ,
for example:

1. <u>Bulletin de l'association des professeurs de
   mathématiques de l'enseignement public</u> (France)

2. <u>Didaktik der Mathematik</u> (West Germany)

3. <u>Educational Studies in Mathematics</u> (Netherlands)

4. <u>L'Enseignement Mathématique, Organe officiel de la
   commission de l'enseignement mathématique</u> (Switzerland)

5. <u>International Journal of Mathematical Education
   in Science and Technology</u> (UK)

6. <u>Mathematics Teacher</u> (USA)

7. <u>Mathematical Gazette</u> (UK)

8. <u>Mathematics Teaching. Bulletin of the Association
   of Teachers of Mathematics.</u> (UK)

9. <u>Mathematik in der Schule</u> (East Germany)
10. <u>Der Mathematikunterricht</u> (West Germany)
11. <u>Zentralblatt für Didaktik der Mathematik</u> (West Germany)

REFERENCES

1. Frederick Mosteller, Chairman, William H. Kruskal.
Richard F. Link, Richard S. Pieters, and Gerald R.
Rising, The Joint Committee on the Curriculum in
Statistics and Probability of the American Statistical
Association and The National Council of Teachers of
Mathematics (editors), a series of 4 books:

    (a) <u>Statistics by Example: Exploring Data</u>
    (b) <u>Statistics by Example: Weighing Chances</u>
    (c) <u>Statistics by Example: Detecting Patterns</u>
    (d) <u>Statistics by Example: Finding Models</u>

Addison-Wesley Publishing Co.,Inc., 2725 Sand Hill Road,
Menlo Park, California, U.S.A. 1973.

2. Judith M. Tanur; Frederick Mosteller, Chairman, William
H. Kruskal, Richard F. Link, Richard S. Pieters, and
Gerald R. Rising, The Joint Committee on the Curriculum
in Statistics and Probability of The American   Sta -
tistical Association and The National Council of
Teachers of Mathematics (editors).
<u>Statistics: A Guide to the Unknown</u>. Holden-Day, Inc.,
500 Sansome Street, San Francisco, California, U.S.A.
1972

CHAIRMAN'S INTRODUCTION

F. MOSTELLER

Harvard University, Cambridge

The past few years have seen a variety of developments
that relate to training in statistics. In the United
States during 1969, we had the Carbondale Conference
(Carbondale, Illinois, USA) on the teaching of proba -
bility and statistics. This was organized by Lennart
Råde and attended by outstanding teachers, probabilists,
and statisticians, several of whom are here today. That
conference produced a book of articles, edited by Råde,
(4) containing useful discussions and descriptions of
programs, and best of all it had many fine examples of
projects and problems useful to the teacher. In my ISI
review (2) of this book I categorized the examples by
type and by article so that they could be readily found.

The ISI has already held two round table conferences on
teaching. The first of these was held at Oisterwijk ,
Netherlands, three years ago. It dealt primarily with
new methods of teaching. That conference produced an
issue of the ISI Review (1) edited by Stuart Hunter,
which had a number of recommendations for training. One
of these dealt with computing and from the Oisterwijk
Round Table grew the Los Angeles Round Table Workshop
on Computing (6). At Oisterwijk it was also suggested
that the training for young people in the important
field of governmental or official statistics was in-
adequate, and yet citizens of every country needed to
consider such statistics almost daily. It was suggested
there that we have a conference on secondary level instruction in statistics, the conference we are now
attending.

In 1972 at the Second International Congress on Mathematical Education it was reported that the British have

now set up a large Committee on the Teaching of Statistics chaired by John Durran, which may review the entire curriculum from kindergarten through the secondary level with a plan to see at what spots in the curriculum statistics and probability should be taught. Also The Open University in England plans a course in probability and statistics starting in 1974-5.

In the United States we have had a Joint Committee on the Curriculum in Statistics and Probability (JCCSP)- joint between the American Statistical Association and the National Council of Teachers of Mathematics-- which I had the honor and pleasure of chairing during its first few years. This Committee has produced two sets of mate - rials whose contents can be judged from the titles and from the one-line descriptions. (See Appendix).

The first set of materials entitled Statistics: A Guide to the Unknown (5) is a set of 44 essays on successful uses of statistics and probability from all the main areas of application, including biology, health, medicine, wild life, anthropology, law, politics, social reforms , language, business, education, governmental statistics, physical sciences, and engineering. These essays are intended to inform people about statistics and about uses of statistical methods. As far as I know this is the first time in the mathematical sciences that we have given a non-mathematical description of the uses of our discipline to the general public. The book is being used extensively in beginning courses in statistics to give a big picture of statistics to people who are just beginning either in mathematical or non-mathematical courses.

The second set of materials is intended to teach sta - tistics. These materials, called Statistics by Example (3), are used to teach elementary statistics by trea - ting real-life problems. And therefore many specific problems, about 50, have been treated at levels approp - riate for students anywhere from ages of 12 or 13 through the early years of college.

Exploring Data shows how to organize data tabularly and graphically and introduces elementary probability.

Weighing Chances develops probability through random num- bers, simulations and simple probability models, and studies more complex data.

Detecting Patterns presents several standard statistical devices-- the normal distribution, the chi-square test , regression.

Finding Models encourages the student to develop models as structures for data so that departures from the model can be observed and new models built.

All four books have teachers' guides including solutions to exercises.

W. H. Kruskal will speak of the future plans of this JCCPS committee now chaired by Richard Pieters. We shall be interested in knowing of related committees in other coun- tries.

This brings us to the Vienna Round Table itself. Why are we here? What can we hope to accomplish?

1. We shall exchange and publish information on what is and what should be going on in statistics around the world at the secondary level.

2. We will review some of the difficulties and needs in the introduction of statistics to school systems and review the differing problems of various countries .

3. We will discuss the role of computers in elementary statistics courses.

4. We will consider recommendations the Conference wishes to make to the statistics profession.

5. We will produce a record of this Conference. Dr. Råde
   will be responsible for editing the work.

6. We will consider whether any long-run program needs
   to be set up by the ISI to improve statistical edu -
   cation at the level of secondary schools.

Let me explain item 6 in more detail. The ISI has two
features of special interest to us. First, it is international, as its name implies. Consequently, it has a
ready line of communication to many countries. If there
were to be an international development in the teaching
of statistics, it would offer one natural home for it.

Second, the ISI has a special interest in governmental
and official statistics. These statistics, although of
great importance to every citizen, have also always been
hard to deal with in school because as they have been
treated in the classroom they seem not to be exciting or
challenging. That this is true is widely agreed, but that
it should be so seems to many of us both unfortunate and
unnecessary. Consequently we raise the question whether
it is possible to prepare materials on official statistics
that are

      i) important to the citizen,
     ii) methodologically instructive,
    iii) intellectually challenging,
     iv) exciting and provocative so that students
          wish to study them.

One question then that the Round Table will deal with is
whether developing such materials on an international
scale would be a wise goal for the ISI, as a long- term
program, and if so what advice we can offer as to its
execution.

Although we hope that other matters will come up as well,
these six points are the key thoughts in the minds of the
arrangers as we open the Conference.

(1) S. Hunter (Ed.). New techniques in statistical teach-
     ing: Report of the International Statistical Institute
     Round Table, Oisterwijk, Netherlands 9-12 September
     1970, Review of the International Statistical Institute,
     Vol. 39, No. 3, 1971.

(2) F. Mosteller. Review of Råde, L. (Ed.). The teaching
     of probability and statistics, Review of the Inter-
     national Statistical Institute (1971), Vol. 39. No. 3,
     pp. 407-408.

(3) F. Mosteller, Chairman, W. H. Kruskal, R. F. Link,
     R. S. Pieters, and G. R. Rising, The Joint Committee
     on the Curriculum in Statistics and Probability of
     The American Statistical Association and The National
     Council of Teachers of Mathematics (editors), a series
     of 4 books and 4 teacher's manuals:

          (a) Statistics by Example: Exploring Data
          (b) Statistics by Example: Weighing Chances
          (c) Statistics by Example: Detecting Patterns
          (d) Statistics by Example: Finding Models

     Addison-Wesley Publishing Co., Inc., 2725 Sand Hill
     Road, Menlo Park, California, U.S.A. 1973.

(4) L. Råde (Ed.). The teaching of probability and statis-
     tics: Proceedings of the first CSMP International
     Conference. Almqvist & Wiksell, Stockholm, Wiley
     Interscience, New York, 1970.

(5) Judith M. Tanur; F. Mosteller, Chairman, W. H. Kruskal,
     R. F. Link, R. S. Pieters, and G. R. Rising, The
     Joint Committee on the Curriculum in Statistics and
     Probability of The American Statistical Association
     and The National Council of Teachers of Mathematics
     (editors). Statistics: A Guide to the Unknown.
     Holden-Day, Inc., 500 Sansome Street, San Francisco,
     California, U.S.A. 1972.

(6) ISI Workshop on Statistical Computation. International
     Statistical Review, Vol. 41, No. 2, August 1973, pp.
     225-275.

DISCUSSION

E. Lunenberg: It is a pleasure for me to welcome the
participants on behalf of the International Statistical
Institute. I want to thank Dr. Schnell and Dr. Leitner
for their stimulating opening addresses and the Federal
Ministry for Education and Arts, and the City of Vienna
for having enabled the Round Table to be held in Vienna.
The Municipal Council for Education of Vienna has kindly

agreed to act as a host for this meeting. I want to ex -
press my appreciation to Dr. Peczar and his staff for
the excellent arrangements they have made.

This Round Table Meeting forms part of the International
Statistical Education Programme of the ISI which was
undertaken under the auspices of UNESCO. In addition to
several other activities in the field of statistical
training and education, Round Table Meetings were held
with the objective to produce reports on various aspects
of the teaching of statistics. At each meeting a small
group of experts has been invited to discuss the topic
assigned to them and to produce (i) a report and a set
of recommendations from which all those concerned with
the teaching of statistics could benefit, (ii) another
set of recommendations to the ISI for possible future
programmes of work in this connexion. The first report
would be published by ISI and widely circulated all over
the world; the second would be submitted to the Statistical Education Committee of the ISI for future action.

I am impressed by the interesting documentation that is
available here as a basis for discussion, thanks to the
efforts of the Programme Chairman, Professor Frederick
Mosteller to whom I want to express my gratitude for all
he has done to prepare this meeting.

A P P E N D I X

Reproduced with the permission of the copyright owners:

Holden-Day, Inc., and Addison-Wesley Publishing Co., Inc.

Statistics: A guide to the Unknown, edited by J.M. Tanur,
and by F. Mosteller, W.H. Kruskal, R.F. Link, R.S. Pieters,
G.R. Rising. The Joint Committee on the Curriculum in Statistics and Probability of the American Statistical
Association and The National Council of Teachers of Mathematics. Holden-Day, Inc., San Francisco, California, 1972.

Part One. MAN IN HIS BIOLOGIC WORLD:

Staying well or Getting Better

The Biggest Public Health Experiment Ever: The 1954 Field
Trial of the Salk Poliomyelitis Vaccine. Paul Meier.
Statistics contributes informed experimental design to
the first attack on polio.

Safety of Anesthetics. Lincoln E. Moses and Frederick
Mosteller. A national study uses sampling and rate
standardization to compare the safety of anesthetics
widely used in surgery.

Drug Screening: The Never-Ending Search for New and Better
Drugs. Charles W. Dunnett. The high cost of searching for

effective new drugs necessitates systematic experimental design.

Setting Dosage Levels. W.J. Dixon. A sequential experimental design reduces the average number of measurements required for a given accuracy in estimating the effects of dosages.

Getting Sick and Dying

Statistics, Scientific Method, and Smoking. B.W. Brown, Jr. The author summarizes the controversy over the effect of smoking on health and explains the value of the debate to medical science.

Deathday and Birthday: An Unexpected Connection. David P. Phillips. Can people postpone natural death? Some famous people seem to have done just that.

Epidemics. M.S. Bartlett. Using measles as an example , this essay supplies both a mathematical model for the rise and fall of an epidemic and a calculation of how small a city must be for the disease to fade out completely.

Men and Animals

Does Inheritance Matter in Disease? D.D. Reid. Comparisons of identical and fraternal twins suggest which diseases are especially influenced by inheritance.

The Plight of the Whales. D.G. Chapman. Methods for es - timating the size of whale populations aid in estab - lishing regulations for their conservation.

The Importance of Being Human. W.W. Howells. A statis - tical technique is used to decide whether a bone fossil belongs to a man or an ape and to help determine how old Man is.

Part Two. MAN IN HIS POLITICAL WORLD:

Governmental Influences on Man

Parking Tickets and Missing Women: Statistics and the Law. Hans Zeisel and Harry Kalven, Jr. Eight examples apply statistics to the practice of law and the study of legal procedures.

Police Manpower Versus Crime. S. James Press. A study in New York City shows that increased police manpower de - creases "outside" crime, but not "inside" crime.

Measuring the Effects of Social Innovations by Means of Time Series. Donald T. Campbell. Data gathered before and after a social reform help to evaluate its effectiveness.

Do Speed Limits Reduce Traffic Accidents? Frank A. Haight. Controlled experiments test the effect of speed limits on traffic-accident rates in Scandinavia.

Statistics: A Guide to the Unknown.

Part Two (continued):

Man Influences Government

Election Night on Television. Richard F. Link. Accurately projecting winners on election night requires rapid procedures for collecting data and a special statistical model to control error.

Opinion Polling in a Democracy. George Gallup. The author illustrates how modern opinion polling gives the citizen a way to communicate his views to the government.

Registration and Voting. Edward R. Tufte. Voting turnout depends on registration, closeness of election, and other social variables.

Part Three. MAN IN HIS SOCIAL WORLD:

Communicating with Others

Deciding Authorship. Frederick Mosteller and David L. Wallace. How statistical analysis can contribute to the resolution of historical questions is shown by an in-vestigation of 12 Federalist papers ascribed to both Alexander Hamilton and James Madison.

Adverbs Multiply Adjectives. Norman Cliff. The "favor-ableness" of an adverb-adjective pair can be quantified as the product of two numbers, one for each word in the pair.

The Meaning of Words. Joseph B. Kruskal. A statistical method uses experimental subjects' judgments of similarity to produce "maps" of words that clarify their meanings on several related dimensions.

The Sizes of Things. Herbert A. Simon. A special method of graphing such things as the sizes of cities and the frequencies of occurence of words in a text produces curves that have much the same shape. The author explores the meaning of these regularities.

Man at Work

How accountants Save Money by Sampling. John Neter. Tests of sampling methods have persuaded railroads and airlines to use samples instead of complete counts to apportion their income from freight and passenger services.

The Use of Subjective Probability Methods in Estimating Demand. Hanns Schwarz. A market researcher tells how to estimate demand for an entirely new product.

Preliminary Evaluation of a New Food Product. Elisabeth Street and Mavis G. Carroll. Surveys and experiments establish the taste appeal and nutritional acceptability of a new food product.

Making Things Right. W. Edwards Deming. Statistical quality control provides an effective means for distinguishing between worker-correctable and management-correctable production troubles.

Man at School and Play

Calibrating College Board Scores. William H. Angoff. Both aptitude- and achievement-test forms must be calibrated in order to make reasonably valid comparisons over time and across subject-matter areas.

Statistics, Sports, and Some Other Things. Robert Hooke. Probability calculations shed some light on when bunts and intentional walks are good strategies in baseball.

Varieties of Military Leadership. Hanan C. Selvin. Statistical analysis shows how even the off-duty behaviour of trainees differs according to the leadership climate within their training company.

Counting Man and His Goods

The Consumer Price Index. Philip J. McCarthy. The author discusses how the CPI is put together and used.

How to Count Better: Using Statistics to Improve the Census. Morris H. Hansen. Sample survey methods have im - proved the accuracy of the census.

Information for the Nation from a Sample Survey. Conrad Taeuber. An associate director explains how the Bureau of the Census determines number of employed and unemployed.

Forecasting Population and the Economy

How Crowded Will We Become? Nathan Keyfitz. The popula - tion cycle has predictable waves that allow forecasting trends and analysis of policies.

Early Warning Signals for the Economy. Geoffrey H. Moore and Julius Shiskin. To forecast the direction of movement of the economy, experts combine selections from among hundreds of changing economic measures.

Statistics for Public Financial Policy. Leonall C. Andersen. Which affects Gross National Product more, changes in money supply or changes in government expenditures?

Measuring Segregation and Inequality

Measuring Racial Integration Potentials. Brian J. L. Berry. A model for measuring the rate of housing desegregation in cities is presented and discussed.

Census Statistics in the Public Service. Philip M. Hauser. Census statistics inform the Chicago Board of Education of the extent of racial segregation in the schools and of steps that might reduce it dramatically.

Measuring Sociopolitical Inequality. Hayward R. Alker,Jr. Measures of inequality can be used to motivate legislative reapportionment and correct racial imbalance in a school system.

Part Four. MAN IN HIS PHYSICAL WORLD:

The States of Nature

Cloud Seeding and Rainmaking. Louis J. Battan. Statistical design and analysis indicate whether Man's efforts at rainmaking actually work.

Looking Through Rocks. F. Chayes. A special method of sampling the surface of microscopically thin sheets of rock produces estimates of the relative volumes of each kind of mineral in the specimen.

The Probability of Rain. Robert G. Miller. By combining the measurements of many meteorological variables from many places, the weatherman computes the chances of fair weather, rain, and snow.

Statistics, the Sun, and the Stars. C. A. Whitney. Correlations between solar brightness measurements have led to a theory of the behaviour of the solar interior.

Modern Machines

Information, Simulation, and Production: Some Applications of Statistics to Computing. Mervin E. Muller. Statistical techniques are applied to both the design and the use of computers.

Striving for Reliability. Gerald J. Lieberman. Reliability theory helps to calculate the chance that a manufactured item will last for a sufficiently long time to perform its function.

Statistics and Probability Applied to Antiaircraft Fire in World War II. E. S. Pearson. The author describes the beginning of operations analysis.

Statistics by Example (in four separate books), edited by F. Mosteller, W. H. Kruskal, R. F. Link, R. S. Pieters, and G. R. Rising. The Joint Committee on the Curriculum in Statistics and Probability of the American Statistical Association and The National Council of Teachers of Mathematics. Addison-Wesley Publishing Co., Menlo Park, Calif.

BOOK ONE. STATISTICS BY EXAMPLE: EXPLORING DATA

1. Organizing Population Data. Douglas Spicer. Organizing data in charts and tables clarifies information about the growth of our population.

2. Fractions on Closing Stock Market Prices. Frederick Mosteller. Exploring data can lead to a theory, but we may need more data to prove or disprove it.

3. A Social Survey of 25 Families. Joseph I. Naus. To summarize our data, we may use averages, and we may also need to get relations between variables.

4. Points and Fouls in Basketball. Albert P. Shulte. If two variables rise and fall together, does a change in one cause a change in the other?

5. Graphical Methods of Studying Data. Yvonne M. M. Bishop. Special graphs called histograms display frequencies; others called scatter diagrams show relations between variables in data about diabetes, cancer, and heart diseases.

6. Babies and Averages. William H. Kruskal. The concept of averages can be both used and abused, as this example shows.

7. Collegiate Football Scores. Frederick Mosteller. Simple questions about scores lead us to organize data into frequency distributions which stimulate new questions. Answering them gives us systematic methods of analyzing data.

8. Ratings of Typewriters. Frederick Mosteller. Sometimes we can get information from data more easily by merely rearranging them in a systematic way.

9. Grocery Prices. William H. Kruskal. We can organize lists of prices to find the most economical way to buy goods.

10. Turning the Tables. Joel E. Cohen. Counting possibilities gives us a way to decide whether an event is as unusual as it seems to be.

11. Testing Beer Tasters. William H. Kruskal. Counting possibilities and multiplying probabilities tells us about the design of an experiment.

12. How Many Fish Are There in a Pond? Samprit Chatterjee. Through the "capture-recapture" method, we can estimate the size of a population that cannot be counted directly.

13. Tom Paine and Social Security. William H. Kruskal and Richard S. Pieters. Two centuries ago Tom Paine made errors, similar to ones we make today, that resulted in underestimating the costs of a proposed welfare program.

14. Fruit Flies. Richard Link. Elementary genetics leads us to develop a binomial distribution.

BOOK TWO. STATISTICS BY EXAMPLE: WEIGHING CHANCES

1. Plagiarism and Probability. William H. Kruskal. An incorrect application of probability.

2. Random Digits and Some of Their Uses. Roger Carlson. What are "random numbers", where do they come from, and how can they be used?

3. How Many Games to Complete a World Series? Martha Zelinka. We illustrate a standard technique for studying statistical processes.

4. Binomial Distributions. Richard Link and Michael L. Brown. A study of the results of picking up pebbles and classifying them as quartzite or not leads to a discussion of the general binomial probability distribution.

5. Black and White Survival in the United States. Bradley Efron. Computing the probability of survival for different groups of people.

6. Introduction to the Chi-Square Procedure. Roger Carlson. Chi-square statistic as a way to see how far observations depart from predictions.

7. Fractions on Closing Stock Market Prices. Frederick Mosteller. Testing to see whether a discrete distribution has equally likely categories.

8. Independence of Amoebas. Joel E. Cohen. Does one kind of amoeba prevent the disease caused by another?

9. What Is the Sample Size? Frank W. Carlborg. Analysis may make clear that the apparent sample size is a highly inflated version of the reality.

10. Rating of Typewriters. Frederick Mosteller. Organizing data in a table so as to make their message clear.

11. Collegiate Football Scores. Frederick Mosteller. The analysis of joint frequency distributions.

12. Periodicities and Moving Averages. Frederick Mosteller. When we use moving averages to analyze time series, we put clear-cut, but artificial, waves into the series.

13. Prediction of Election Results from Early Returns. Joseph Sedransk. How proper adjustment of partial information permits accurate early predictions.

14. The last Revolutionary Soldier. A. Ross Eckler. Applying statistical analysis to the historical problem of establishing a date.

BOOK THREE. STATISTICS BY EXAMPLE: DETECTING PATTERNS

1. Predicting the Outcome of the World Series. Richard
    Brown. By direct mathematical attack and by com -
    puter simulation, we calculate the probabilities of
    the overall outcome of a sequence of independent
    events.

2. The Normal Probability Distribution. Roger Carlson.
    The normal probability distribution is discussed and
    its use as an approximation to discrete distributions
    explained. Tables and graphs of the normal are used.

3. How much Does a 40 Pound Box of Bananas Weigh? Ralph
    D'Agostino. Bananas differ in weight, so boxes cannot
    be packed to exact weight; also they shrink during
    shipping. How much variation in packing weight can
    be tolerated?

4. Grocery Shopping and the Central Limit Theorem. Samuel
    Zahl. Central limit theorem is applied to a house -
    wife's method of quickly estimating cost to show if
    she buys several items her results will probably be
    satisfactory.

5. Chi-Square Distributions by Computer Simulation. R.K.
    Tsutakawa. Chi-square test is applied to two examples:
    the recovery of banded geese, and the effect of cloud
    seeding on rainfall. Simulations of sampling by use
    of random numbers is used to study the properties of
    the chi-square statistic.

6. Sensitive Fingers and Defective TV Tubes. Ivor Francis
    and William H. Kruskal. In testing extra-sensory per-
    ception, production of TV tubes, or social theories,
    the hypergeometric probability distribution may aid us.

7. Studying an Expert. Ralph D'Agostino. Ideas of con -
    ditional probability lead to an analysis of errors in
    a long series of forecasts of crop yields.

8. The Case of the Vanishing Women Jurors. Stephen E.
    Fienberg. Although over 50% of possible jurors are
    women, Dr. Spock at his famous trial faced an all -
    male jury. The author studies the probability of this
    using the binomial and hypergeometric distributions.

9. Transformations for Linearity. Frederick Mosteller.
    Complicated relationships between two variables may
    be easily determined by graphical methods if trans-
    formation of one variable leads to a straight line
    graph. The method is applied to earth vibrations near
    waterfalls and to the distribution of primes.

10. The Acceleration of Gravity. John Mandel. To bring
    order into the data of many experiments, the author
    fits straight lines to a large set of data and thus
    gets estimates of the acceleration of gravity.

9. High Energy Elementary Particle Physics Experiments. L. Keith Sisterson. Particles collide and produce new kinds of particles. Properties of the Poisson help physicists figure out which products appear after collision.

10. Remote Geologic Mapping. Paul Switzer. Data obtained from airplane runs over rocky terrain give us probabilistic information about the kind of rock in the territory. The author discusses how to use it.

11. Estimating Population Sizes. Janet Wittes. From several incomplete lists of members of a population, we want to estimate the size of the population. Through a statistical model a method is found and illustrated.

12. How to Buy a Used Telescope. A. Ross Eckler. A problem in how to make the best buy of a used telescope is solved by maximizing a probability which itself is obtained as a sum of conditional probabilities. Applications of the method to other but similar problems are considered.

TEACHING OFFICIAL STATISTICS

B. BENJAMIN

Civil Service College, London

1. Introduction

Before discussing the teaching of official statistics
there are two immediate questions to be answered (1)
what do we mean by official statistics? (2) to whom is
our teaching to be directed?

The impression given by the sections of general statistics textbooks which are devoted to official statistics
is that these are very largely descriptive and historical.
They describe the social and economic conditions of the
population as they were at the last population census;
they describe the working of the economy in the recent
past - the total national product, the terms of trade ,
the level of investment; they include various serial statistical indicators - of the cost of living, of wages ,
of production.

This view of official statistics is incomplete, increasingly so. It ignores the very important change in the
utilisation of statistics and in the role of the statistician in government which has taken place progressively
in the last two or three decades. The volume of routine
statistics produced by government departments has not diminished. On the contrary it has increased both in cover-
age and detail. Indeed the volume is so great that it has
become increasingly necessary for government statistical
offices to provide guides and indexes to help the user to
find his way to the particular elements of the overall
measurement that interest him. The much more important
change that has occurred in addition to this growth is

that the administrator and the manager have been educated[1]
to appreciate the role of information in decision-taking.

The administrators who advise the elected representatives
in government now have a greater understanding of the
social and economic objectives of government, of the need
to construct strategies to achieve these objectives, and,
more particularly, of the need for information to illumi-
nate this process of policy-making. This understanding is
becoming communicated to the elected representatives.

The New Approach

It is important to spell out what this change means. It
means that the administrator or politician must henceforth
expect to be articulate about objectives and strategic
problems; to be prepared to express his problems in quantitative terms. He must be prepared to translate the
language associated with his management situation into
terms which the statistician can, in turn, convert to
information requirements. We stress here that "information
requirement" is radically different from "data collection".
Information is the analysis and assembly of undifferenti-
ated data into the answers relevant to the question asked
by the administrator. Everything depends, therefore, on
the administrator having sufficient perception of his
information requirements to enable him, albeit with the
help of the statistician, to frame these questions. This
done, the manager must be able to trust his statistical
advisers to produce objective answers. This implies that
he has chosen them for their competence and professional
approach.

1)
This education has been partly formal, resulting from the
growth of management and business schools, staff colleges
and the like; partly informal and stimulated by hard events-
economic crises and wars - which highlight strategic de-
ficiencies.

This does not mean that his job is to be done for him by his statistical advisers. It is the administrator or politician who makes the judgments and the decisions. For his own job satisfaction the statistician will want to explain the strategic implications of the decisions, and be listened to. For _his_ job satisfaction the administrator will want to retain his own freedom of movement and the right to do "his own thing" even if this includes trusting his own intuition and doing something which is electorally rather than statistically acceptable.

It can therefore be seen that in the light of this change the whole concept of official statistics has to be revised. Official statistics are not to be seen only as accounts of transactions, activities and services though these are undoubtedly indispensable, partly as, in themselves, direct answers to policy questions and partly as source data for the derivation of specific answers to specific questions; we must extend the definition of official statistics to the multifactorial analysis of social and economic data, to the study of relationships and the isolation of controlling and controllable variables, to the construction of statistical models for predicting the probable consequences of action. The definition must embrace a much more sophisticated and more determined attack upon uncertainty than contemplated or hitherto understood by official statisticians.

Correspondingly the statistical machinery of government both central and local is now increased in scale and importance and in the extent of penetration of the policy-making arena. It has its own problems of administration and therefore more understanding of administra - tive behaviour; but, more important, it confronts many more administrators. The need for administrators to come to terms with it is now inescapable.

Training of the Statistician

Let us now turn to consider those who are to be educated;
and first, the statisticians concerned with official statistics. It is no longer good enough as it may have been
in the past, for the official statistician to be a descriptive economist with a fair understanding of index numbers
and not much else. We now demand a sound first degree
knowledge of statistical theory; or if the first degree is
not in statistics as the principal subject, a post-graduate
diploma or master's degree in statistics. In addition (and
not in substitution) a knowledge of economies; of the
machinery of government; and last but not least a practical
knowledge of the difficulties and limitations of data
collection and processing in the central and local government context. This post-graduate knowledge is of such a
practical nature that it cannot but be acquired after en-
try into the public service and government must be pre -
pared to provide it by organised means. It is not going
to be good enough to rely on "sitting by Nelly" for this
is leaving it too much to chance both in opportunity for
learning (Nelly may be too remote for some) and scope of
training (some Nellies are more communicative than others).

So in relation to the important training in methods of
data collection and assessment of reliability and the
proper use of data processing resources, the best Nellies
must, together with the junior statisticians be brought
to the classroom as a formal uniform and thorough exer -
cise. There, too, must be given short courses in economics,
government machinery, and for the benefit of the later
responsibilities of these junior statisticians some instruction in the basic principles of good management.

There is one further requirement. There is now a wide
variety of statistical tasks in the public service, some
involving a different range of technique and different
knowledge of data sources from others. It is important

to keep the challenge of difficulty and innovative spirit alive by not allowing statisticians to stay too long in any one task location; if this happens the statistician tends eventually to accept a routine and to cease to develop; to acquire a vested interest in the status quo. At this stage separation of man and job can be traumatic. The post-entry education of the statistician, therefore, never ceases but involves continued movement at fairly regular intervals from field to field. The intervals should not be too short as to damage productivity, nor too long to cramp development. A central statistical service must provide for career planning and do more than pay lip service to it.

The Administrator

Formidable as this demand for the education of the statistician may be, the educational need of the administrator is a much larger and more difficult to satisfy if only because the basic numeracy which he requires (and to which reference will be made later) ought to be inculcated at a very early age.

The objectives are:

1. to provide the administrator with an appreciation of the way in which information can be and is being used in the government service to assist administrative decisions;

2. to improve recognition by administrator of the quantitative elements within any problem with which he is faced; to help him to perceive when he needs statistical assistance;

3. to enable the administrator to express the quantitative parts of his problems in numerate terms; so as to help the statistician to recognise not only the implicit information needs but also the form in which those needs should be satisfied in order that the decision-making may be best assisted;

4. to provide the administrator with sufficient appreciation of the tools which the statistician possesses
to deal with situations of uncertainty and estimation,
that he has confidence in the competence and objectivity of the statistical services available to him;

5. to enable the administrator to understand better
the limitations as well as the full implications of
the results of investigations carried out by the
statisticians.

In order to achieve objective (1), and to some extent
also objective (2), it is essential for the teaching to
be carried out clearly within the Government setting.
Teaching illustrated by reference to industry, commerce
or the natural sciences is not effective for administrators; indeed it tends to create antipathy. There are
exceptions in certain specific situations; for example,
the basic concepts of probability can be easily conveyed
by reference to games of chance. But where a problem of
decision is to be the vehicle for demonstrating some
technique, the problem must be recognised as one which
they are likely to encounter in their working experience.
The reason is simple though not obvious except to those
with some experience of adult education. It is that once
formal education at school or university has ended, we are
all of us less able and less willing to abstract methodology from one situation in order to apply it to some
other situation. It is necessary to make some concession
to this difficulty by arranging to demonstrate a wide
variety of official problems so that there is an en -
hanced likelihood that something will be recognised as
relevant by everyone.

Where a wide variety of problems is presented, the
opportunity should be taken to demonstrate the fundamental universality of the basic statistical method; to show
that while techniques are specific to particular situation,

(and the <u>statistician</u> must master them) statistics, basically is not all about techniques but about the simple
concepts of classification and comparison.

Having convinced the administrator by relevant illustration, that information (which is nearly always statistical) does come and should come, into Government decision-
making, we have next to get him alerted to recognise
that information may also be essential to effective decision-making within his own sphere. The approach here
is to ask administrators to consider any problem, however trivial, in their current experience which may have
some statistical content. These problems can then be
discussed by other administrators under direction. The
discipline of thinking of these examples has in itself
a marked effect in sharpening awareness of the statistical nature of problems which had hitherto not been seen
in this light. It has an advantage too in enabling those
who are responsible for teaching to get a closer under -
standing of the real day-to-day needs of administrators
in relation to level of appreciation, depth of knowledge,
and perception of the relative roles of the administrator and the professional statistician.

The next step is to help the administrator actually to
express the quantitative nature of his problems in terms
which (1) are articulate to him so that he can see, himself, what his information needs are and (especially important) the way in which the information will eventually
be used to solve the problem and (2) are articulate to
the statistician who will be called upon to organise the
data collection and analysis. The statistician will need
especially to know how the information is going to be
used in order to advise upon the most economical scale
of data collection and in order to appreciate the format
in which he must prepare his final analyses and report.

Basic Concepts

This means teaching the administrators the primary
elements of the language of statistics; the basic concepts of classification and comparison and the sub -
sidary concepts of frequency distributions, measures of
central tendency and dispersion, probability and correlation and even some idea of decision theory. There are
also some derived concepts like index-numbers and time-
series which are of particular importance to administrators. These concepts must be described as they are but
they will not be retained unless they are also conveyed
within case-studies recognisably of relevance to the
Government scene, which involve them. The concepts can
then be emphasised, parenthetically, without digressing
too far from the problem and without the digressions
being resisted as text-book teaching. It is much more
effective to introduce the concept of a frequency distribution in examining the general problem of unemployment (length of time unemployed) than to look at the
life-times of electric light bulbs. It would have more
meaning to compare two age distributions if this were
against the background of the problem of planning social
welfare services.

The case-studies referred to here are not the three-day
marathons beloved of business schools. Short, half-day,
expositions, with participation by the administrators,
are likely to be more attractive as well as more effective.

One must not, however, be either too casual or too subtle
about the introduction of theory if objective (4) is to
be achieved. If we are to give administrators some confidence in the sophisticated tools that statisticians
use, we must at least identify and describe the occasions
of their use. The aim is to use appropriate opportunities
during practical illustrations to show, without going

into theoretical details, how the statisticians can command knowledge of probability and draw on exper - ince of the behaviour of economic or social statistics to derive relevant and usable information from situa - tions, more the rule than the exception, in which data are insufficient either in quantity or specificity. In this context, "usable information" means information on the basis of which a confident decision can be taken, the degree of confidence (or reliability) be - ing either (1) specified in advance by the administra- tor (thereby determining the amount of data to be collec- ted) or (2) specified by the statistician as the best that can be got from the most efficient use of the li - mited data available. Alternative (2) is the most that can be done within the constraints of time and money .

We have to avoid two dangers. First, if we make these techniques look to complicated either by going into too much mathematical detail or by over-emphasising pitfalls, then the administrator may close his ears and eyes to them altogether or he may regard them as too "clever" to be trusted. Instinct turns us all toward the simple solutions. Second, if we over-simplify these techniques then administrators may be unwittingly encouraged to collect them in "cookbook" form and use them indis - criminately without understanding the conditions in which they are appropriate or their limitations. Some economists and many engineers do this. At best it can lead to the frustration of obviously nonsensical re- sults; at worst it can lead to wrong decisions and costly failures.

Computers

Because there is a danger that a computer can come to be regarded as an end in itself rather than as a means

to an end, there is a need to avoid teaching adminis -
trators about computers as if it were a subject in
itself. It has to be emphasised that a computer cannot analyse anything; it can only carry out analysis
that have already been humanly designed; that a bad
survey is not converted to a good one because as the
newspapers so often impress upon us, "the result will
be fed into a computer". Undoubtedly computers do take
a lot of the chore out of statistical analysis but it
is what the statistician is trying to do that is important. In itself the computer is no more important
than the typewriter. It is therefore preferable to
reserve only a short period for explaining the way in
which computers work, how they are programmed and what
sort of tasks they are best at doing, and for the rest,
to refer to computers and the way in which they facilitate statistical analysis as part of the description
of those exercises and their methodology.

Ends and Means

The same distinction between ends and means is important
in relation to information system, so-called. An in -
formation system makes no contribution to policy-making
unless the policy-maker uses information; otherwise, it
may be an expensive white elephant. One does see many
systems which are well designed as systems for storage
and retrieval, communication, and feed-back which make
little reference to the nature of the information and
hardly any reference to the way in which the manager
may use the information and the way in which the system
may therefore be interrogated. Such systems may be fois-
ted on an unsuspecting administrator by an enthusiastic
computer manager. So in teaching administrators, it is
better to refrain as far as possible from speaking of
management information systems as a particular subject
but rather to concentrate upon the need for information

in decision-taking. It is desirable that a management
information system should not be presented as a theoretical model but a working model within the Govern -
ment environment that can be seen to be contributing
to managerial capability. A system that could be used
is unimpressive; administrators are willing to accept
only examples of managerial tools that, demonstrably,
are working and are worth the effort and cost involved.

Basic Quantitative Thinking

We come now to the most basic ingredient of all - the
initial level of ability, on the part of the administrator to think quantitatively. It is very difficult
to begin to teach an appreciation of official statistics to an administrator who finds it almost impossible
to think in terms of quantities, to compare quantities
and to order them in magnitude. This is sometimes referred to as being innumerate. Numeracy is however a
much abused word in this context. Its use tends to
suggest and indeed does mislead those whom we are trying to teach into thinking that we are concerned with
dexterity in arithmetic or the use of algebra as a
language. It is probable also that many academic teachers
of statistics would regard this kind of numeracy as the
inertia barrier beyond which it is necessary to have
their students initially carried. It might be so for
statistics students but it is not so for administrators
(or ordinary members of the public). (Those who teach
administrators should not be academic in their approach).
It is not of primary importance that administrators
should initially be good at arithmetic or able to recognise that $y = dx + b$ is the equation of a straight line
with slope a and intercept b. It _helps_ but it is not
essential to their understanding of the power of the
statistical method or of the value of information or of
the way information should be used.

The simple ideas they need initially have to do with the association of like with like; the separation of homogeneous groups (leading to notions of sets); relative sizes; differences within and between groups extending to the concept of variability (even if only vaguely held). Given this, all the rest that we have described can be, in sufficient measure, added.

These ideas ought to have been acquired at school , especially in an era of "modern" mathematics, but arts graduates are still entering the public service without having acquired them. For some time therefore it may be necessary to provide remedial treatment before the main body of the training may be attempted. This has to be handled within great skill. We are remedying a deficiency of childhood education. It involves treating adults like children. The condition for success are (1) that the treatment must be selective - they must classify themselves; (2) it must be frankly admitted that the treatment is juvenile in the best sense (not in the sense of treating them as mentally deficient). The "best sense" meaning that many parts of the adult mental furniture are acquired as juveniles; (3) it must be stressed that it is not a matter of experience and not a matter of being clever or not clever; that it is all well within their ability; (4) the treatment must be handled by deeply perceptive and lucid teachers with a professional (but not an academic) approach.

Secondary level of education

It is possible to be more specific about these statistical ideas which we would like to introduce and to have absorbed at the secondary level. The approach should be somewhat as follows:

We can begin by reminding pupils that numerical facts
of some kind are rarely absent from their daily lives,
in government, industry, commerce and even in leisure;
that they have only to open the daily newspaper or
watch television to find some reference to changes in
income tax, to the balance of imports over exports (or
vice versa), to prices, to traffic on the roads, to
index numbers of production or to the volume of scales
of some new product, and to cricket or football scores
or the odds paid on winners of horse races. They are
presented with, for them, the very real situation that
the untrained human brain tends to baulk at large numbers and especially at a collection of several numbers.
They can be told that the training required to facilitate their comprehension involves (apart from arithmetic
and other forms of mathematics, i.e. a knowledge of how
numbers themselves behave) the development of a method
of analysing and assembling these numerical facts in
such a way as to render them intelligible; that this
method is called the statistical method or more often
statistics.

Inference

Statistics, they can be told, enable us to describe what
we have observed, but the statistical method enables us
to go further and to draw, from past observations, inferences about likely future events; that it is this
capacity for making the world around us more ordered,
and therefore easier to come to terms with, that
makes the statistical method such a powerful tool. Many
of the day-to-day decisions we make are based upon the
regularity of behaviour of observed happenings in the
past. When we stand at the bus stop we are acting on
our knowledge of the frequency of buses in the past;
when we "post early for Christmas" we are heeding a
warning based on the observed frequency of postings at

previous Christmases; when mother bakes a cake she uses
a recipe that has always turned out well in the past ;
the fuel we lay in for winter is based upon a general
expectation of the amount of cold weather that is like -
ly to be experienced; the shopkeeper orders from the
wholesaler on the basis of recent demand (together with
expected response to current advertisment) for various
commodities; when the merchant buys barley in the market
he not only accepts the handful of corn from the sack
which the farmer has brought as indicating the quality
of all the corn harvested from a particular field but
he also compares this with other samples already offered
to him by other farmers on the same or previous market
days (later the children can be reminded that he is, in
most cases unconsciously, making use of much modern statistical theory - sampling technique and the comparison
of group averages); when the weather forecaster issues
a gale warning to shipping he is assuming that an atmos-
pheric disturbance will move and develop in the same way
as similar past disturbances the behaviour of which has
already been charted; when we work out tactics for an
important football match we do so on assumptions, based
on experience, about the behaviour of intelligent players
in different circumstances.

Variance

We go on to say that they may occasionally hear someone
say " it is impossible for me to reduce my past experience
to numbers, since every circumstance is quite different
from another". For example, a schoolteacher may say this
of his pupils and the doctor may say it of his patients.
Actually variability (variance in the language of statistical measurement) is an essential part indeed the
foundation of statistical theory. The statistician learns
to recognise similarities between members of a group and
to use a measure of likeness, i.e. of tendency to conform

to type (central tendency) as a starting point for reducing differences to a mathematically describable
pattern of deviation from a defined type (just as in
coordinate geometry, points in a curve are referred to
an origin). Likenesses and differences are the raw
material of statistics. The statistician does not say
"all men are equal but some are more equal than others";
he says "all men are different, but some are less different than others".

We may then offer an example. If we examine a large number of classes each of 40 students we shall find most
often that 4 are lefthanded writers and 36 righthanded.
Any two classes may be quite different in this respect;
one may have no lefthand writers, another may have seven.
If the chance of any child becoming a lefthanded writer
is one in ten (we shall have to say later what we mean
by chance), then in a large number (say 1,000) of classes
each of 40 we should expect to find the following incidence of lefthandedness:

Frequency of occurrence of different
numbers of lefthand writers in 1,000
classes of 40 pupils each

Number of left-hand writers	Occurrences
0	15
1	65
2	142
3	200
4	206
5	164
6	107
7 - 9	96
10 or more	5
	1,000

We then indicate that this kind of table is called a
frequency distribution. It shows how the 1,000 classes
are distributed among the different possible kinds of
event (0, 1, 2, .. lefthand writers). This calculation
makes it possible for us to accept that the risk of
lefthandness is one in ten without expressing surprise
when occasionally (in 10.7 per cent of occasion) we
find 6 in a class of 40. But if we found 10 lefthand
writers we would regard this a very rare event; so
much so that we might suspect one of the important
underlying assumptions of our calculations, namely,
that all the classes are representative of the general
population of children ("unbiased samples"). We should
say that the difference between an actual frequency of
10 and an expectation of 4 is statistically significant.
This is another way of saying that we are suspicious of
this particular class and that we have to take into
account the possibility that this is no ordinary class.
When we use the term "statistically significant" we are
only making a statement about numbers; there is still a
question to be answered - "significant of what?". A
possibility to be investigated is that this is a class
formed in some way that tends to bring together lefthand writers (this sample is biased in the direction of
increasing the number of lefthand writers). If for
example some areas of the country had a higher incidence
of lefthand writing than others (if lefthand writing
were correlated with area of residence) then the words
"in some way" used above might mean that the pupils
came from an area of high incidence.

Confidence intervals

When we are indicating the likelihood of different numbers of lefthand writers in classes of 40 pupils we
would not write out the table though it would be more
informative to do so. As a shorter method of describing

the situation we should indicate the limits beyond which
we should be suspicious of the representativeness of the
class. In ordinary conversation they would hear people
say "I should be surprised if there is less than 1 or
more than 8". Statisticians say much the same thing in
a slightly different way. They say they are confident
that the number will be between these limits. Certain
conventions are adopted to indicate the strength of this
confidence.

We talk about 95 per cent confidence limits meaning that
these limits will (in representative samples) embrace
95 out of a 100 occasions and will only be exceeded in
5 occasions out of a 100. Alternatively, we refer to
the range between the limits as a confidence interval.
The odds against a value outside these limits turning
up is 20 to 1 (strictly 19 to 1) and this is regarded
as such heavy odds that if it does happen we regard it as
statistically significant and want to probe for special
factors.

One might then go on to talk about statistical significance and estimation more fully but as far as possible in
everyday language. We could introduce the ideas of a
random process and of bias. We can talk a bit about the
design of experiments and the analysis of variance.

Having reviewed these ideas which have been underlined,
it is possible to review some of the practical problems
which statisticians have to deal with and in which ideas
are involved.

Then we must talk about some of the techniques involved
just as a carpenter might show off some of his tools and
explain what particular jobs they are used for (without
imparting skills to his listeners). Sampling, classification, comparison, errors of observation, frequency dis -
tribution, measures of central tendency and of dispersion,
histograms, come into this discussion. Classification for
comparison would be the dominant theme.

The next stage is to deal with background mathematics
and especially probability. As far as possible one
ought to ensure that the following theorems and procedures have been covered:

> Arithmetic and geometric series. Indices and
> logarithms. Simultaneous linear equations. Fitting
> of straight lines and simple curves by least
> squares. Permutation and combination. Binomial
> expansion. Exponential and logarithmic functions.
> Differentation and integration of polynominals.
> Maxima and minima. Area under a curve.

However it is my view that these help but are not essential for progress in the absorption of the basic statistical concepts I have referred to. Probability theory at
an elementary level _is_ essential however and will be a
matter of central interest.

Again one would refer to the intervention of chance and
likelihood in everyday life (crossing the road for example
or life assurance). One might talk about games of chance
and the frequency concept of probability. Equally one
would introduce a geometric approach with Venn diagrams.
There would be plenty of examples and problems drawn
from practical situations. It should be impressed on
them that chance, error, and inference come into official
administrative decisions. They must not be allowed to think
that chance is an element of importance only in games
and puzzles. The more that they know about chance and
likelihood before leaving school the better. It is a
good deal easier to understand at 13-15 than at 23-25
and certainly much easier than at 33-35.

D I S C U S S I O N

F. Mosteller: In any future text material we prepare,
we will want to catch the students' interest. We have
not worked hard in the past on capturing the attention
of girls in our statistical materials. Consequently we
ought to review methods of e.g. doing advance market
research so that a reasonable fraction of the examples
and exercises have interest for women. Naturally the
ideal problem interests everyone, but such problems are
rare in real life.

H. Aigner: If adults are innumerate, some reasons may be
a) plain forgetting, b) years of experience may have
passed without any need for numeracy, c) a cultural bias
against quantitative as opposed to qualitative thinking.
Difficulties in remedying this situation at school include a) failure to make a clear distinction between
teaching an appreciation of statistics and teaching statistical methods up to an acceptable level of skill ,
b) inability to forecast the needs of students at the
secondary level (unless they are in a technical school),
c) difficulty of finding applications that are close to
reality, appealing to students and at the same time not
too time-consuming.

IDEAS FOR TEACHING OFFICIAL STATISTICS

S. A. GOLDBERG

New York

Introduction

1. Two different ways of using the word "statistics"
   a) connotes theories and techniques dealing with
      two main questions:
      (i) methods of procuring information;
      (ii) methods of describing and analysing informa-
      tion. (Theory of probability, sampling, correlation,
      tests of significance, averages, frequency, dis -
      tributions, etc. included under a)).
   b) A synonym for "information" expressed in numerical
      form. In this form information is -
      (i) relatively precise, though may be approximate;
      (ii) e.g. if I say there are many people in this
      room - What is "many" - 10? 50? 100? 500?
      But if I take a count and say there are 40, I may
      still be mistaken but, as a rule, close - there may
      be 35 or 40 or 45, but not usually 50.

2. When we talk about official statistics we deal with
   b) primarily, though techniques of a) utilized.

3. Official statistics frequently regarded as cold and
   impersonal. There are good and valid reasons for this
   which may appear evident later. However, it is para -
   doxical that this should be so because much of the
   information deals with the most intimate and important
   happenings of the individual, his family and his environment. This is illustrated in main part of document.

4. Approach used: tell story of man and his social and
   economic life in a series of major sequences, high-
   lighting them with important illustrative social and

economic concerns. Each sequence can be a chapter or
a portion of a chapter. Appendix to each chapter
could deal with sources and methods, relating to
particular sequence, and some illustrative statistical material. Sections can then follow on usefulness of statistics and some characteristics of national statistical offices - the usual sources of
official statistics.

5. The story of man, despite its vast diversity, can be
told under six main parts:

Part 1   Man (man embracing woman, as usual)

Part 2   The institutions of man and his relation -
ship to them. Three (or four) classes of
institutions may be distinguished:
(1) the family
(2) business institutions
(3) government and private non-commercial
institutions.

Part 3   Goods and services that man produces, buys,
sells or gives away either directly or
through institutions with which he is associated (including harmful things - environmental
pollution);

Part 4   Natural resources (and natural gifts such as
fresh water, air, etc.)

Part 5   Man in the national setting

Part 6   Man in the international setting.

In what follows a few illustrations are given:

PART 1 OF THE STORY: MAN

1. Begin with beginning - the birth of a human being. This
is the conventional beginning of official statistics.
(There is a prior beginning reflected in Professor
Kinsey's pioneering work; also, reference to evolving
fertility surveys and surveys of family planning).

Source: in many countries registration of birth is compulsory and documents used for statistical purposes. Also censuses and sample surveys.

2. <u>As people grow older</u>, statistical agencies keep track of their advancing ages - through censuses , surveys, administrative records, estimates, etc.

3. As individuals grow up various things happen to them which have a public interest and hence form subjects for official statistics:

    (a) number and characteristics of individuals attending <u>school</u>, colleges, etc. Full-time, part-time, vocational, on the job training, etc.

    (b) <u>education</u> is one of fortunate things. There are also unfortunate things in which there is a public interest. One of these is getting into difficulties with the legal authorities. Classification of crimes. Sources of information - often court records and special surveys - e.g. of vic - tims surveys. Problems of definition, coverage and international comparisons.

    (c) Another unfortunate event - illness. Classification of - problems of definitions - degrees of illness and health, etc. People in hospitals and at home. Sources of information.

    (d) It is man's lot, soon or later to have to earn his livelihood. For most of us this means "work" as it has been ordained: "In the sweat of your face shall you eat bread until you return into the ground". Incidentally, official statistics cover information on number and characteristics of <u>those who return</u>. Aside from those who return - and therefore are absent from the labour force - there are those who choose not to enter it, or to leave it, for various reasons - the sick, the retired, the voluntarily idle. The dilemma of the housewife who works but is not in the "labour force".

(Sources - registrations, labour force surveys, censuses, administrative records, etc.)

(e) The time comes when most of us choose to get _married_, though bachelors (spinsters) would claim that this only happens to those with least resistance. (Data on marriages collected in same manner as statistics on birth).

PART TWO OF THE STORY: INSTITUTIONS OF MAN

1. Individuals are, of course, part of a _family_ before and after marriage. With marriage a new family is formed. Characteristics of families - composition , size, income, expenditure. Censuses, surveys. Also separations, divorces, etc.

2. Individual in course of his life becomes part of other _institutions_.
   a) He may get a job in a _business organization._ He becomes a member of a different institution - as he becomes an employee, manager or owner. Characteristics of business organizations - various fields (agriculture, manufacturing, transportation services, etc. Sources of information censuses and sample surveys. Also tax data.
   b) _Government institutions_ - growing importance and diversity. Characteristics and functions reflected in statistics. Sources of data administrative re - cords and surveys.
   c) Other non-commercial institutions - Some hospitals, schools, charities, etc.

PART 3 OF THE STORY: GOODS AND SERVICES OF MAN

1. Created by man in his institutional setting and they have an existence of their own. Illustrations of

commodities and services produced, exported, imported.
Description of sources.

2. Characteristics of goods and services - e.g. prices.
Price is a creation of man - he creates it through
market mechanism of supply and demand, but once created it may be regarded as a characteristic of the
things of which price applie's. Index numbers, etc.
Sources of information.

PART 4: NATURAL RESOURCES OF MAN

Often not measured or measurable: air, water, mineral
deposits, land, parks, etc.
Environmental pollution - the antithesis of production
of Part 3.

PART 5: MAN IN THE NATIONAL SETTING

Section 1. The National Economy and its parts
The production, distribution and financing of the nation's output of goods and services involves countless
transactions - buying and selling goods and services;
hiring labour; investing capital; paying taxes; and so
on. The records of these transaction form the basis of
producing numerous statistical series, as just indicated above. However, in order to provide a manageable
picture of the structure of the economy and its func-
tioning, these series must be summarized into a limited
number of significant categories. The most comprehensive of these summaries are:
1 - the national income and expenditures accounts
     which provide an essential quantitative frame -
     work for studying the state of the economy.
2 - input-output tables which are designed to show
     the sales of specific industries to all other

industries and the consumption by specified industries of goods and services produced by other in -
dustries.

3 - <u>financial flow accounts</u> which provide information
on changes in assets and liabilities of the various sectors of the economy and are designed to show,
among other things, who is borrowing and who is
lending and the financial instruments through which
this borrowing and lending takes place.

4 - <u>Balance of International Payments</u> - giving as a
picture of our transactions with foreigners.

5 - <u>national production and productivity indexes</u>, for
which statistics are arranged in an entirely different way.

Section 2. <u>Social Indicators and Social Concerns as reflected in statistics</u>
A description of current work in this sphere na -
tionally with some selected illustrations.

PART 6: MAN IN THE INTERNATIONAL SETTING

1) The nation, the continental regions, the world.
2) The continental regions and world as extension of nation, through trade and other transactions, migration,
communication etc., reflected in international statistics.
3) The heterogeneity of world (rich, poor, advanced, backward,) reflected in international statistics.
4) The homogeneity and interdependence of regions and
world reflected in international statistics.
5) International work in social indicators.
6) Selection of major international social concerns.

The Usefulness of Official Statistics

Part A   By Analogy
Part B   The emphasis on interrelationships

Part A. USEFULNESS OF STATISTICS BY ANALOGY

I. Statistics must have purpose and fulfill urgent needs-
   otherwise there would be no point collecting them.
   a) demands for statistics have become greater and grea-
      ter - why? (access to good and timely information
      puts a country (business) in better position in a
      competitive world. Knowledge has come to be recog-
      nized as an important factor of production and offi-
      cial statistics are a vital element of this know -
      ledge; marked broadening of economic and social goals,
      and desire for growth and stability, nationally and
      internationally; the increasing utilization of com-
      puter making possible all types of retrievals not
      possible previously, etc.)
      1) official statistics and economic and social man-
         agement and planning;
      2) official statistics and business management and
         planning;
      3) official statistics and economic development;
      4) official statistics and business accounts - ana-
         logy.

II. The statistical system may be likened to a mirror -
   a) Why do we use a mirror? In part it is, of course, in
      order to experience the feeling of joy in seeing our
      faces.
      Statistics, too, may be used to indicate achieve -
      ments - of a country, a state, an industry, or the
      average family, and make people feel good.
   b) But, that aside, we look at a mirror in order to get
      guidance for decisions and action.
      1) when we see our hair mussed up - this is a signal
         that we should comb it;
      2) when we see bags under our eyes this may concern us-
         we stop to consider what caused this and come to
         the conclusion that we are working too hard.

In the same way statistics are a basis for action -
1) when gross national product is seen to be moving
   in a certain way; or employment and unemployment
   developing in a certain way; or when certain
   diseases become prevalent - then some decision may
   be forthcoming to change, encourage or do nothing
   about the situation.
2) the decision may deal with administrative matters,
   as when wages are tied to the consumer price index;
   or when the distribution of seats in Parliament is
   adjusted according to figures of the population
   censuses; or when the contributions of a country
   to the United Nations financial budget is based on
   its per capita national income.

c) Analogy with mirror may be carried further;
   1) How well we see ourselves when we look in the
      mirror depends, in part, on the distance at which
      we stand. If we stand far from the mirror we get
      an overall view of ourselves, but we do not detect
      details. As we approach the mirror, the details
      become more pronounced. But when we approach too
      close we may get a lopsided view of ourselves. The
      question of balance and perspective come in.
   2) This is also true of statistics. Many statistical
      series deal with aggregates. They are overall mea-
      sures. Their purpose is to shed light on the over-
      all contours of an economic or social problem.
   3) However, to fulfill this purpose it is also necess-
      ary to study comparable data for the various fa-
      cets of problems.
   4) Thus to understand the overall picture we must
      study a considerable amount of detail - (that is,
      statistical breakdowns, for example, by age, mental
      studies, region, income, industry, etc.); and to
      understand the detail properly we must have an over-
      all view. If we go too far in the direction of detail

we may lose perspective - moreover, it may become too
costly (in the same way that too great a concentration
on the details of our face in the mirror may be too
costly in that it may leave us heartbroken). We have
here, too, a question of balance and perspective, depending on the purpose for which we want the statistics.

d) A statistical system may be likened to a mirror from
still another point of view.
  1) A mirror may be so constructed as to give us a
  distorted view of the image we see. Similarly, a
  well-painted face may appear in the mirror much
  better than it really is;
  2) This is also true of statistics. They may be abused.
  They may be employed to sensationalize, confuse,
  over-simplify, (See How to Lie with Statistics by
  Darrel Huff.)

e) Another aspect of the analogy:
  1) the image in the mirror is seen better and more
  realistically against a background of proper lighting.
  2) this is also true in using statistics - a problem
  can be seen better if it is studied against the
  background of related data.

f) The analogy with the mirror breaks down in two important respects:
  1) First, each image in a mirror is a fleeting event-
  when you leave the mirror the image disappears. By
  contrast, statistics can accumulate, giving us the
  opportunity of making comparisons with the past.
  Statistics when they are available over a period
  of years, become elements of economic history -
  a vehicle for better understanding the present and
  the emerging future. By summarizing certain aspects
  of the experience of the past, statistics help to
  establish probabilities for the future.

2) Second, when we look in a mirror the object of our
preoccupation is a particular individual image.
By contrast, official statistics deal, not with individuals as such but with populations, totals, the
mass in which the individual, as such, is entirely
submerged. Statistical aggregates, averages, frequency distributions, and so on that Statistical
Offices publish are abstractions, and they reveal
characteristics not necessarily about any component
individual but about the group as a whole - which
gives us the key to the question raised before why
official statistics are considered cold and im -
personal.

Part B.   THE USEFULNESS OF STATISTICS AND INTERRELATIONSHIP
          OF SOCIAL AND ECONOMIC PHENOMENA

1) This could be made into a most interesting section by describing how statistics are used to assess the interrelationship between various social and economic changes, e.g. ,
between changes in fertility and income, education, occupation, age; the relationship between size of income, education, occupation areas of residence, etc.; relation bet -
ween death rates and illnesses and income, geography ,
occupation, etc; relationship between employment and un -
employment and education, income, mobility, etc.; and so
on.
2) The implications of interrelationships on policy forma -
tion, both private and government.
3) Quantitative Interrelationships, model building and the
Computer.

SOME CHARACTERISTICS OF AN OFFICIAL STATISTICAL AGENCY

I.   It would be interesting to say something about the
organizations representing the immediate source of official
statistics.

a) The most important characteristic is co-operation –
a spirit of co-operation must pervade an organiza –
tion providing good statistics. This necessity for
co-operation stems from the very nature of official
statistics.

   1) How do official statistics come into being? In the
last analysis they come into being by processing,
adding together and analyzing information which
the public supplies the statistical office. Thus
production of good statistics is ultimately de –
pendent on co-operation and goodwill of those who
supply the basic data – individuals, business
organizations, hospitals, governments, etc. (Just
to make sure co-operation is forthcoming we have
legislation making it mandatory; the opposite
side of this coin is secrecy provisions of statistical legislation – the issues of invasion of
privacy and statistics).

b) Official statistical production is a practical man's
preoccupation – its purpose is to produce a basis for
action (or research).
Co-operation with users of the statistics is essential
in order to keep abreast of and understand their needs.

c) Statistics are, as a rule, used in conjunction with
one another – not in isolation (reference to inter –
relationships). Statistics in related fields must
therefore be comparable. This means statisticians must
co-operate with one-another (nationally and internationally).

II. Characteristics of people working in statistical offices'

a) subject matter specialists – must know field, understand needs of users, problems of suppliers of data
and be able to communicate with them and appraise
statistical results for reasonableness and relevance.

b) survey (mathematical) statisticians for designing
and conducting sample surveys, censuses, etc., and
evaluating results.

c) technicians and clerks.

III. <u>Research and research tools in a statistical office</u>
a) Classification systems (of commodities, industries,
diseases, etc.)

b) development of new measurable concepts;

c) determining reliability of results.

d) development of synthetic statistics - indicators,
indexes, seasonal adjustment, etc.

e) analysis of results, etc.

IV. The computer technology and statistical availability
and analysis.

OFFICIAL STATISTICS

OF POPULATION AND PRODUCTION

(Themes for Teaching at the Secondary Level)

NATHAN KEYFITZ

Harvard University, Cambridge

Isolated Numbers and Ratios.

That the population of Amarillo, Texas was 127,000, that
58,000 women were employed gainfully in Trinidad, that
192,000 deaths occurred in Romania, are all official statistics for the year 1970, and they could well be true
statements. The first might interest the city treasurer of
Amarillo because it receives a subsidy per head of population as counted in the census, but otherwise such isolated
numbers do not satisfy important preoccupations. They take
on somewhat more interest when compared with other numbers:
the employed women can be presented as a percent of all
women; the deaths can be broken down by age and referred to
exposures to risk. From the latter we could go on to derive
probabilities for individuals, for example a life table
giving probabilities of survivorship age by age, and these
lead to expectations of life for individuals and groups.
Ratios and probabilities are a first step in putting numbers
to work, but they are only a first step.

Past Changes and Extrapolation to the Future.

The population of Amarillo or any other count can be com -
pared with a similar count made a decade earlier. Such
first differences over time help us to interpret totals and
ratios: we usually care less about population, employment of
women, or probability of dying at a particular date, than we
do about the change from one date to the next. First differences of time series implicitly bear on the future, and the

hint they provide of what may happen in the coming year or
decade causes readers to turn to official statistics with
avidity. They note, for example, that crimes rose only 7
percent between 1970 and 1971 in the United States, whereas
they had risen 11 percent on the average during the 1960s,
and this suggests that they may soon stop increasing alto -
gether and start to decline.

We want more and better official statistics, and we want
them as comparable as possible between successive years,
just so as to make implicit or explicit extrapolations. The
most urgent demand for official statistical data is in
business and governmental planning. They permit currents
of trade and migration to be traced, labor surplus areas
to be identified, future shortages of goods to be antici -
pated. No one can afford to make a decision to expand an
industrial plant, to start a new line of production, or to
curtail an existing operation, without a picture of what is
already going on, and official statistics contribute impor-
tantly to that picture.

But a depressing feature of these applications is their
rate of obsolescence. Once a year has passed and we know
the official 1973 production or other figures we forget
the crude extrapolation by which we anticipated it. The
official 1973 number is more accurate than the best fore-
casts from 1972 and earlier years can be. Regarded this way,
official statistics, like the daily newspaper, become ob -
solete very soon after they are published. This paper will
show how statistics raise questions and permit conclusions
whose validity and interest go beyond the year to which
they refer and beyond their immediate use by business and
government.

Statistical Description of Urban Processes.

Statistical tabulations may refer to entities that are of
purely formal definition and with not much bearing on

social dynamics in the field in question. In respect of
urban population it was seen half a century ago that the
incorporated city was not the right entity to tabulate for
most purposes because suburban settlements functionally related to the city had become numerically significant. Once
many people with jobs in the central city moved out to a
suburb, tabulations for the central city ceased to satisfy,
and a changeover was made to Metropolitan Areas of cities.

The journey to work --how far people commute-- has now become of interest and has been incorporated in the U.S. Census questionnaire. Surveys have become capable of measuring
the daily pulsation of labor force back and forth between
the city and its suburbs, and so provide a framework for
explaining changes in social habits like the shortened lunch
hour and the possibly imminent 4-day 40 hour week. And this
daily and weekly cycle is superimposed on a new kind of life
cycle for the individual family of postindustrial society,
by which the young couple start life in an apartment in the
central city, move to the suburbs when they have children;
twenty years later their children leave home; the suburban
house and lawn are more trouble than they are worth, so the
couple move back to the central city. They now have the income to occupy a high-rise, high-rent apartment, and to enjoy expensive amenities of the central city. This life cycle,
clearly indicated by a suitable assembly of U.S. official
statistics, will be found in Europe and elsewhere with the
spread of industrial society.

Thus official statistics can firm up and delineate with
some precision a picture that we may have vaguely in our
heads before we look at them. The statistics take on im -
portance according to,the importance of the processes they
describe. As the processes and problems change, so must the
statistics. Rural-urban migration, important for over a
century, is necessarily coming to an end, while intra-urban
and interurban movement is more rapid than in the past.

The urban entity is also changing. Cities in the old sense,
incorporated as well as metropolitan areas, seem to be
losing some of their importance to much larger built-up
areas, from Boston to Washington on the east coast, from
Milwaukee through Chicago and past Gary at the south end of
Lake Michigan. Compilations for these larger areas will be
increasingly needed, along with data on migration among and
between them. In the nation as a whole people seem to be
leaving the interior and moving towards the coast; the 1972
Statistical Abstract (page 6) of the United States Bureau
of the Census for the first time showed the population within
50 miles of an ocean; it turned out to be 54 percent of the
total, following a steady rise from 46 percent in 1940.

Old and New Problems of the Economy.

The art of the official statistician, which is to see or
sense what is happening in his society and to devise statistics that will describe it, has especially wide scope
in the economic field. In the 1930s and 1940s the economy
became an important preoccupation and economic series of
many kinds were developed more or less independently --
prices, carloadings, wages, profits. They badly needed co-
ordination. This was accomplished through the device of
the national accounts, by which a variety of transport,
production, wage, and other data were cleared of duplica-
tion and otherwise adapted so that they could be added up
to the gross national product and national income. The
national income series of the 1930s and onward portrayed
the shift of production from the household to the market,
along with the increase in market production. The expansion of restaurants, laundries and other commercial services
that do what was formerly done in the home is a process
that may be called monetization of the economy. Production,
investment, wages, profits and taxes are readily definable
elements of the monetized economy and can be shown with
precision in the national accounts.

Since Keynes the accounts have been used by governments to
control the economy, and thereby serve, at least in prin -
ciple, as part of a feedback system hopefully providing
stability. Without the Keynesian picture of the world the
national accounts as we know them need not have been in -
vented.

But in recent years attention has moved beyond total and
average of money income. Equality has become important; we
want to know how the national income is distributed as well
as how much of it is produced. The poor as well as the rich
indeed benefit from rising national income per head, but
poverty nevertheless stubbornly persists, and the relation
of poverty to total income has become problematic. Simple
frequency distributions in terms of current dollars grossly
exaggerate the diminution of poverty; 1947 showed 27 per -
cent of families with incomes less than 2,000 Dollars in
the United States, while 1970 showed only 5 percent less
than 2,000 Dollars. To allow for inflation and other factors the Bureau of the Census calculates families on low
income, suitably defined so as to be genuinely comparable
from year to year, and finds the number to be 39 million
in 1959 and 25 million in 1970. (Statistical Abstract, page
329). Poverty comparison between nations as well as through
time require not only statistics of income, but also of
factors related to the behaviour of lower income groups,
for instance their morale. One suspects that when and if
such are produced they will not show the richest country
to have the least poverty.

Distribution of income as between the sectors has also become important, especially between government and the
private sector. The share of government represented by the
fraction of national income originating with it has gone
up in the United States from 9 percent in 1950 to 14 per -
cent in 1970. (Abstract, 1972, p. 317) This is net of
such government payments as unemployment compensation and
old age pensions, which are called transfers and are properly excluded from the national income. But such payments

and the taxes that support them are clearly needed for estimating the increased power of government. For this the en -
tire budget is pertinent, and we note that the Federal budget alone exceeds 20 percent of the national income. Power,
like an effective international comparison of poverty, takes
us past the frontier of present data. It may be a concept
beyond any possible response of official statistics.

The Environment.

A different kind of challenge to official statistics is the
limit set on human activity and progress by the physical
character of the planet Earth. Our generation perceives population and economic growth as pressing on finite stocks
of fossil fuels and metals, with limited air and water to
carry away such undesired products of production and consumption as sulphur dioxide, unused fertilizer and insecti-
cide, and scrapped automobiles. The national accounts treat
the petroleum in the ground as of zero value to the comm -
unity --they add up payments, all made to people; no pay -
ment is made to nature for putting the petroleum in the
ground during the course of millions of years. Yet only
a finite amount of petroleum was deposited. Production in
the 48 States has averaged 6 percent per year increase during this century; to extrapolate into the future at 6 per -
cent would be very wrong, for oil production is near its all
time high and likely to decline.

Attention properly shifts to reserves of fuels and metals,
and statistics of these reserves, however uncertain they
may be, take on new importance. Once exploration for a
given fuel or metal ceases to offset mining withdrawals we
are on the down slope of the production curve. To know just
where we are on that curve tells how hard conservation must
be pressed, substitutes sought, new technologies developed.
If things go badly, statistics of air pollution will in the
future become comparable in importance with statistics of
automobile production. Insofar as pollution is the result of
the automobile and also a limiting factor on its use, data

on it is needed to coordinate transport policy.

Work and Leisure.

The solution of economic problems shifts the distribution of people's activity. In particular, as productivity increases some of its benefits are taken out in leisure. One would expect hours worked in advanced countries to show a steady decline, as in recent years in the United States: average weekly hours in manufacturing dropped from 41.2 in 1965 to 39.9 in 1971. (Abstract, 1972, p. 225) But the long-term trend is less clear; the 1950 figure was 40.5 . Perhaps people do not want free time in the degree that increased productivity would permit them to have it. Manu - facturing output per man-hour rose by 81 percent from 1950 to 1971 (Abstract, 1972, p. 232), while hours worked dropped by only 1 1/2 percent. Age at retirement is falling, but some of this is forced retirement, and so analogous to un- employment. People complain about unemployment even when their incomes are maintained. Perhaps work gives people a kind of satisfaction that they cannot quite have from leisure. Can we find a measure of pleasantness or unpleasant- ness of work to go along with established measures of quantity? To tell why work is so important to people we have to go beyond the frontier of official statistics.

We do have extensive statistics on leisure activities. U.S. recreation expenditures went from 11 billion Dollars in 1950 to 39 billion Dollars in 1970, an increase far beyond that of population and prices. Attendance at major league baseball went from 18 million to 29 million in the same period. Golfers who played 15 rounds or more per year, only 3.2 million in 1950, were 9.7 million in 1970. Whether such numbers can ever be assembled into a time budget that would help trace the evolution of post-industrial man remains to be seen.

Health Services Input and Length of Life Output.

Official statistics are useful if they point to an important
and difficult problem even when they cannot yet throw much
light on it. A further instance is the relation between
death rates and medical services. One would have thought
that as the amount of medical services goes up death rates
would go down. A glance at official statistics for the
United States shows the relation to be far from simple.
Health expenditures went from 13 billion Dollars in 1950
to 71 billion Dollars in 1970, a multiplication by 5.6 .
During the same time population increases in the ratio 1.34,
and consumer prices in the ratio 1.61, so in real terms per
capita health services multiplied by 2.58, still a large
increase. The year 1950 showed medical expenditures at
4.0% of GNP, 1970 showed 7.4% (Abstract, 1972, p. 65). Over
the same 20 years the expectation of life at age zero went
up from 68.2 to 70.8, a very small increase, especially compared with the decades of the 1920s and the 1940s, in each
of which the expectation of life rose by more than 5 years.

Because some of the rise in medical expenditures is due to
better pay for medical personnel we turn to measures of
service, and find physicians (233,000 in 1950 and 348,000
in 1970) and active registered nurses (375,000 in 1950 and
700,000 in 1970) both substantially increasing, while
hospital aides and orderlies went up even more. Hospital
beds on the other hand increased less rapidly than population, and, in fact, declined from 9.6 beds per thousand
population to 7.9 (Abstract, 1972, p. 71). One would have
thought effectiveness of use of hospital beds is increasing,
but the statistics include no measure of this. Specializa-
tion of physicians and group practice, the elimination of
house calls, all point to more intensive use of medical
personnel. By 1970 only 58,000 of the 334,000 physicians
in the United States were in general practice. (Abstract,
1972, p. 68).

Additional medical attention presumably brings diminishing
returns in longevity, so perhaps all we can expect for the
158 percent increase in deflated expenditures per head is the
further 2.6 years in the average expectation of life shown.
But what is genuinely puzzling is how unequal is the diminution of the death rate through the several age-sex categories.
Female mortality fell at all ages, and male mortality some -
what at younger ages, but virtually no improvement was shown
for males beyond age 50. In the age group 65-74 the U.S.
death rate was 48.34 per thousand in 1945-49 and 48.60 in
1960-64.

Not all of the expenditure on health is intended to reduce
the death rate. The expenditure includes cosmetic surgery,
dentistry, and a variety of thoroughly worthwhile treatments
designed to ease pain. But can we not assume that some fraction of health services expenditure ought to act on the death
rate, and when we find that the death rates go down for some
age-sex groups and not at all for others, ought we not to
be surprised at the unequal impact? Some influences on the
death rate like penicillin are of trifling cost. Smoking,
drinking, air pollution, obesity, lack of exercise, and
other consequences of high consumption, all act on the death
rate; have they offset increased medical expenditures?

Existing official statistics at least put us in a position
to state the problem of medical policy in its most general
form: how can the nation's health budget be allocated to
minimize sickness and death? Present statistics provide
some but far from all of the means to tackle this question.

Traffic Mortality.

Automobile accidents present some special issues within the
field of mortality. The number of fatalities in the United
States went up from 34,763 in 1950 to 53,041 in 1966. When
we relate these to the population series, we find that in
1950 the deaths were 23.1 per 100,000 population, and in
1967 26.7, still a rise but much smaller. But is not the

proper denominator for automobile accident fatalities the amount driven rather than the population? Taken in rela - tion to vehicle miles which had just about doubled between 1950 and 1967 we find an actual decline, from 8 per 100 million vehicle miles to less than 6. (Vehicle-miles, cal- culated from gasoline sales, are the best measure of ex - posure available; it would be better yet to use person- miles if this were known.) Apparently the long-term de - cline is due to the changed pattern of driving -- especi- ally that more of it was on newly constructed freeways which despite appearances are safer per mile than the slower driving on rural and urban roads.

More difficult to understand are the personal factors in accidents. In 1967 white males showed 41.1 fatalities per 100,000 population, and white females 14.4 or about one third as many, both on an age-adjusted basis. Are women more careful drivers? Do they drive less? We have no record of the amount driven by men and women. Do they drink less? Even less do we have a record of the amount of drunken driv‌- ing by men and by women. Do they do their driving at hours when there is less traffic, and hence when the roads are safer? Without answers to such questions the sex differences in accident rates are not understandable.

Both instructive and baffling are the differences between marital status groups. Married men have the lowest accident death rates, with 36 per 100,000 population; single are next with 56; then divorced with 122; finally widowed with 127. The single are, on the average, younger, but this has been met in the above rates by an adjustment for age differ- ences. Among possible causes of the large differences are married persons driving less, driving more carefully, drink- ing less before setting out, or driving better cars. On none of these factors do we have data. That married men are healthier is suggested by their lower death rates from all causes; perhaps they also have better eyesight. That widowed

and divorced men are over three times as liable to be vic-
tims of automobile accidents (through their own fault or
that of others) is of interest for its own sake and prac-
tically valuable for insurance underwriting. That our
knowledge goes as far as the above observations is both
an achievement and a challenge to official statistics.

Agriculture and Probability.

Random variation is a stubborn hindrance to proper interpretation of official statistics, for example in respect
of agricultural production. Consider two headlines in the
New York Times. "Food Production Gaining in India; Dream
of Self-Sufficiency Believed Near Reality" (September 6,
1971), and 17 months later, "India Sees Tough Times;
Prices Rise, Income Lags" (February 23, 1973). The latter
article was based on a decline in food production in the
preceding year, the former on an increase.

The series of five good harvests, climaxing in 108 million
tons in 1970-71, provoked the euphoria of September 1971.
Then followed two bad years, and it looks as though the
1972-73 crop will be less than 100 million tons. Extrapolation from two or five years in the series, subject as
it is to random variation, is bound to lead to alternation
of un warranted gloom and enthusiasm. Such cycles of des-
pair and hope go back many generations into Indian history.
A longer perspective would avoid them and permit more effective policies, though it would deprive us of spectacular
headlines.

One way to see what is happening is to portray population
on the horizontal axis of a plane and grain production on
the vertical, with each crop year represented by one point.
The question at any time is wether population or food is
winning out, and to decide which we need to establish one
line at a suitable distance above and one below the ir -
regular points of the data. Statistical theory is well

suited to the probability interpretation of official time series.

Measuring the Green Revolution.

Underneath the random process we find today the systematic effect of the green revolution. A simple model will suggest the relation between higher yields and population. Suppose that new rice varieties, row planting and massive application of fertilizer can double the yield in a given area, and therefore ultimately in the whole of a country. Suppose that farmers holding 5 percent of the nation's land effectively adopt the methods each year. This is fairly rapid diffu - sion, considering the conservatism of peasants, the limited number of agronomists to instruct them, problems in the adaptation of the new techniques to various soils. At the end of 20 years the country's total crop will be double that of the starting time.

Suppose also that the population is increasing at 3.3 percent per year--this is more rapid than India's present growth, but population of some other countries are increasing faster than 3.3 percent. Then at the end of 20 years the national population has doubled. Per capita availability of food grains is exactly as at the beginning.

How can population increase of 3.3 percent per year keep up with a rate of diffusion of the new methods at 5 percent per year? The answer is that the 3.3 percent is a geometric increase, based at each moment on the population currently attained, while the 5 percent is essentially arithmetic increase, based on an initially given area.

One doubling is not the limit of agricultural advance. Further developments in seeds, and more massive applications of fertilizer are all possible. A second doubling would make yields per acre four times what they are now. But if the second doubling requires a further 20 years, then the same effect is repeated.

Such gradual improvements do nothing but increase the scale
of the problem. Measures like outmigration that would have
been possible on the small scale become impossible on the
larger scale.

The green revolution speeds the monetizing of the economy.
The farmer has to borrow for his inputs, and to repay the
loans he has to sell much of the crop. In some instances
the small Javanese farmer sells the crop standing in the
field to a buyer who has his own hired work force to cut it,
a radical departure from traditional practices. The farmer
is richer, but he has become aware of the cost of maintaining his relatives, service caste affiliates, and inefficient
laborers. For these the green revolution makes access to the
crop more difficult. If they can find jobs in the city, all
will be well; the success of the green revolution depends on
the expansion of industry no less than of agriculture.

This sketch of possible demographic and social consequences
of the green revolution suggests that statistics of grain
yields and population growth are a bare beginning in trac -
ing its effect. Statistics are needed to show the spatial
diffusion of new agricultural methods, monetization and dis-
placement of traditional work styles that follow on this
diffusion, migration to the city and availability of work
there, if anyone is to know in what degree increased yield
makes better lives and in what degree it enlarges the scale
of poverty.

Preparation of Instructional Material.

Official statistics can only be taught, at the secondary or
any other level, in relation to problems, practical or theoretical. To teach the numbers and definitions as such, without relation to problems, would be to imply that the statis-
tics-collecting activity goes on for its own sake.

The preceding remarks have sketched a few of the kinds of
issues that give life to statistical data. To regard a
metropolitan area as pulsating in a daily rhythm, and the

individuals in it as going through life cycles that take
them out to the suburbs in their twenties and bring them
back to the center in their fifties; to regard health
services as input in a system of which lowered morbidity
and mortality are the output; to see the agricultural
aspects of the green revolution in conjunction with its
population aspects; all these can become teaching materials
only with much further examination, documentation, and
simplification.

Exercises of various kinds are needed to help fix the
material in the student's mind. He could be asked to do
a report on the relation of health services to death rates
in his own country, insofar as indications for both are
given in his official reports. From the same source he
could examine trends in death rates according to cause,
and show how infectious diseases have diminished, while
degenerative ones have increased. But these are too sophisticated for starting on; prior to them he should be asked
simply to look up the population, deaths, and national income, and calculate the death rate and per capita income.
He needs practice in searching and interpreting official
primary sources.

In short, this paper is a long distance from a textbook.
At best it suggests how questions may be juxtaposed to
official statistics to give them meaning. I believe it
feasible to create materials that will teach the sources
of official data, and obliquely to introduce the student
both to statistical theory and to social science.

STOCHASTICS AND MODERN ALL - ROUND EDUCATION

W. EBERL

Technische Hochschule, Vienna

Stochastics as a unit.

"Stochastics" is a comprehensive name for a large variety
of branches of pure and applied mathematics whose unify -
ing link is the notion of probability. The term stems
from the Greek "στοχαδεσιβ" which means "to aim, to
guess, to suppose" and is already found in modern textbooks, periodicals and reference books. Some German universities have changed the name of their institutes of
statistics calling them now institutes of stochastics.
There is also a periodical "Stochastics" (2). Statistics
is only a part, important as it may be, of stochastics
which is connected with many other branches of stochas -
tics.

The basis of stochastics is probability theory. Since 1933,
when A.N. Kolmogorov laid down its axiomatic foundations on
the basis of previous work of R. von Mises, probability
theory has become a part of measure theory, distinguished
by its intuitive interpretation and its practical applicability. In the middle of this century the impact of topology and functional analysis stimulated the development of
the theory of stochastic processes in the framework of probability theory. Stochastic processes may serve as mathe -
matical models of empirical processes influenced by chance.
They are employed in the natural sciences, in the technical
and medical sciences, and also in economics, e.g. in forecasting.

If in slight extension of the usual meaning an experiment
is defined as an action that may be repeated very often
and the outcome of which is more or less uncertain, then
games of chance, measurements and statistical surveys are

also experiments. In a large majority of cases the application of probability theory to problems of an empirical na - ture consists in the selection of a stochastic law or model on the basis of a finite number of outcomes of a certain experiment. This procedure is called inductive inference and is subject to a greater or smaller uncertainty. With the help of probability theory statistics give a mathematical form to inductive inference so that it may be performed with a numerically fixed degree of certainty. Mathematical statistics may therefore be defined as the mathematical theory of inductive inference.

In so far as the observed data are often the result of a long series of measurements or of statistical surveys, statistics may also be defined as the mathematical theory of mass phenomena influenced by chance.

The latest development of statistics is motivated by the ideas of game theory (3). Game theory originates from in - vestigations of strategic games like draughts or chess and constructs mathematical models of the rational behaviour of persons or firms, that are involved in economic conflict situations, e.g. competition. This theory also considers games in which, as in card-playing, the outcome of a game depends not only on the strategies of the players but also on chance. Game theory therefore makes an extensive use of stochastic notions and methods. Statistical decision theory (4) transfers the basic idea of game theory to statistics, by interpreting inductive inference as a game of the statistician against nature, in which the statistician by his decision has to minimize the value of suitably chosen loss functions. Many important methods of statistics are in this way only special cases of a general theory of optimal de - cisions in the face of uncertainty.

The branches of stochastics consist in the majority of cases of methods that have been developed for a certain field of applications. Some of them are:

Characteristic functions, correlation theory, design

and analysis of experiments, ergodic theory, estimation
theory, information theory, Markov processes, martingale
theory, Monte-Carlo methods, multivariate analysis, non-
parametric statistics, probability theory, queuing
theory, regression theory, reliability theory, sampling
theory, sequential analysis, stationary processes, statistical decision theory, statistical quality control ,
stochastic differential equations, stochastic integrals,
test theory, variance analysis.

This enumeration is far from complete. Between these branches there are many connections and overlapping and there
are comprehensive monographs about each of them. As the notion of probability is common to all of these subjects, it
seems natural to put this vital mathematical organism, called
stochastics, into the field of public attention and interest
and to incorporate some of its basic ideas into the system
of secondary instructions.

All-round education.

Education should help a person to master the difficulties
of life, to make his life richer and finer, and to become
a useful member of society. Education is always multiform.
Not only is there a distinction made between scientific,
artistic, and political education, but also education of
the character and of the heart. All-round education is the
harmonious union of all these components in the unity of a
personality. Education does not consist in the accumulation
of knowledge and abilities, but in the readiness and in -
clination to make practical use of what one has learned.
Or as Goethe said: "Everything is hateful to me that only
instructs me without widening or stimulating my activities."
If one would test the curricula of our secondary schools
on the basis of Goethe's statement, e.g. by a statistical
inquiry of working adults who have successfully completed
their secondary studies, considerable time and energy could
be saved in favour of subjects that are of more general
usefulness.

The aim and purpose of education, namely the maintenance
and improvement of life, are timeless. But the actual subjects of instruction depend largely on the day and age.
What claims are to be made on education in an industrial
society? What is the part of stochastics in it?

Industrial society.

The industrial society is " a society whose essential features stem from technology and economics, especially from
mechanical production and its effects on human individuals
and communities" (1). According to ideological foundations
and political systems several variants of industrial society may be distinguished. But some characteristics are
common to all of them.

The members of the industrial society enjoy a relatively
great security of existence and a high standard of living,
unequalled till now. Differences of classes have diminished
to a large extent and an economic or social rise for most
of the people is possible. In general this requires a high
degree of personal achievement and a persistent zeal for
learning. People who have worked their way up are over -
burdened with work and are under constant pressure.

By the rapid development of modern systems of transport
and communication, by technical and economic interweaving
of far distant areas, and the mass production of good and
cheap non-fiction books, people of our time are confronted
with an excessive supply of information of which it is
only possible to receive and assimilate a very small part.

In the pluralism of the democratic version of industrial
society the technical sciences which have their origin in
the cooperation of mathematics and the natural sciences
gain more and more reputation and weight, because they form
the foundation of this convenient way of life and because
they show by the incontestable objectivity of their results
a pleasant and striking distinction from the variegated
diversity of popular opinions. The cooperation of mathe -

matical and empirical research has made promising progress
in the field of economics where statistics (in the frame -
work of econometrics) is one of the main tools of pioneering work.

The loss of authority of traditional systems and institutions
corresponds to a vigorous effort for a rational understand -
ing of reality. Statistics as a theory of inductive inference
plays a leading part in this endeavour.

Masses of different kinds are very frequently found in in -
dustrial society. Urbanization leads to overcrowding, the
high standard of life is based on mass-production and mass-
consumption, information is spread by mass-media, traffic
and holidays degenerate to mass-traffic and mass-tourism.

Here are the practical tasks and possibilities of stochastics
and statistics. Statistical quality control comprises, as the
name indicates, control procedures that ensure a satisfying
quality of mass-produced goods in spite of the largely anonym-
ous manner of production. Queuing theory supplies optimiza -
tion methods in cases when a mass of "customers" (e.g. travel-
lers, telephone subscribers, engines in need of service or
repair) has to be "serviced" in an economical way. Inventory
theory makes it possible in many cases to determine optimal
inventory sizes if supply and demand are stochastic. Sampling
theory gives rules, how to test a mass of similar objects
economically by samples if a certain reliability of the test
is prescribed. The list could be continued a long while, but
it is evident, that stochastics and statistics are of eminent
importance for a rational handling of many widespread prob -
lems in an industrial society.

Stochastics in school.

A school system that aims to prepare young people for life
in industrial society has, therefore, to incorporate the
basic ideas of stochastics, especially of statistics, in its
curricula. The following points should be observed:

It is impossible to enlarge subjects of instruction without reducing others. The inclusion of stochastics will there – fore meet with strong opposition on the part of all those who are hit by the reductions. If a general agreement about the omission of time-honoured subjects cannot be reached, there remains only the approved procedure to establish and offer different types of curricula and to incorporate stochastics in some of them. Certainly it is possible to facilitate this development by an international exchange of experiences and by informational support by mass-media.

The selection of subjects from stochastics for use in secondary schools calls for particular care. One should pay attention to the following demands:

a) Limitation to simple ideas that have an intuitive background and

b) that can be handled exactly with mathematical methods at the pupil's disposal.

c) Emphasis on the connection between statistics and electronic computers which are indispensable for a rapid evaluation of extensive numerical data.

d) Illustration of the wide applicability of sto – chastic methods with examples from different fields of great general importance e.g. (public health, environmental protection, traffic, economics).

e) Emphasis on the fundamental part of statistics as a mathematical theory of inductive inference.

To introduce stochastics in secondary education teachers and textbooks are required.

It is therefore necessary to make stochastics obligatory in the curricula of teachers' training colleges. Textbooks should be written according to rather concrete recommendations of a board of experienced teachers and experts on stochastics.

Prospects.

The result of an integration of stochastics, in particular statistics, in secondary education should be a new attitude in relation to problems of our time: people whose education is strongly influenced by stochastics have fewer prejudices and more sound realism. They are more inclined to reject dogmas, ideologies and slogans, as soon and in so far as they become aware of them.

If possible,they like to make decisions on the basis of reliable numerical data. But they are realistic enough to know the limitation of such a quantitatively oriented way of thinking.

References

(1)   Keilhacker, M.: Erziehung und Bildung in der indus -
      triellen Gesellschaft. Kohlhammer, Stuttgart, 1967

(2)   Tintner, G. (ed.): Stochastics. Gordon & Breach, Lon-
      don.

(3)   von Neumann, J. and Morgenstern, O.: Theory of Games
      and Economic Behaviour. University Press, Princeton,
      1944.

(4)   Wald, A.: Statistical Decision Functions. Wiley, New
      York, 1950.

DISCUSSION

F. Mosteller: I would like to ask whether stochastics will include the new codification of exploratory data analysis as developed by J.W. Tukey and his colleagues. This material has the property of being on the one hand a set of mathematical devices, but with no probability actually showing and with only modest mathematics being used, the functions being linear, quadratic, powers, logarithms and so on. One difficulty is that the mathematics teacher may not be willing to treat such methods since the attention is heavily oriented to the applications. The computer can be used to reduce the time required for the more tedious aspects of the work.

W. Kruskal: Recent developments in exploratory data analysis
suggest a great variety of deep, difficult problems in mathematical statistics: namely, to understand the stochastic behaviour of these data - exploratory methods under various
kinds of assumptions about the underlying structure. For
example, there is little knowledge of the behaviour of re -
latively simple clustering algorithms under what one might
think are simple assumptions about structure.
These same recent developments in data exploration - developments largely stemming from J.W. Tukey's work - may be re -
garded as a revival of serious concern for often divided
metthologies that have not been thought about with much care
in recent years. For example, graphical methods, tabular
methods, and analysis of variance in the sense that some of
its practitioners adopt.
Exploratory data analysis has led to considerable controversy, both about its role in statistics as a profession, and
perhaps even more about its role in statistical education.
Perhaps the most basic question is that of dangers in putting
with professorial approval of recipes into the hands of stu -
dents who may well not appreciate difficulties, traps, the
importance of framing the right questions, etc. One can think
of obvious, if perhaps unfair, analogies from the training of
surgeons.
At least one British statistician, A. Ehrenberg, is interested in data analysis for official statistics, thus combining two themes we have discussed. He is writing a book on
the subject.

B. Benjamin: What worries me about "data analysis" is the
risk that you may convey the wrong impression of the way
in which statistical methods ought to be used. Many of us
who have been practising statisticians all our lives have
somewhat bitter memories of earlier days when administrators
came to us with voluminous data with a request to "make something of it". All too often we had to reply that nothing
could be made of it because the data were derived from asking
the wrong questions; that the design of the enquiry was the
all important part of the statistical method not the data
analysis; that they should have come to us earlier so that
we could have advised them about the kind of data they
really needed; that the data analysis itself could often
be very much simplified by cares in question design or
choice of measurements. The frustration of this situation
may very well create an antipathy to the statistical method.
A worse danger is that a busy statistician may be tempted to
give the enquirer a "cook-book" to enable him to perform data
analysis himself. This is highly risky and may often lead to
the production of nonsense results. This too tarnishes the
reputation of statisticians and impedes the advancement of
the subject.

R. Rao: While I am in a broad agreement with the ideas expressed in Dr. Eberl's paper, I do not favour the proposal of teaching statistics as a separate subject at the high school level. We should aim at rationalizing teaching of other subjects by introducing statistical ideas and quantitative thinking in discussing the problems related to particular subjects. In such a case we will not be teaching statistics at the expense of other important subjects like mathematics, physics etc. Some of the topics mentioned in Dr. Eberl's paper cannot be introduced at the high school level as the students may not have the requisite mathematical equipment to understand them or to learn them in a rigorous way.

Reprint from STATISTICS AT THE SCHOOL LEVEL
Published by Almqvist & Wiksell International, Stockholm

COMPUTING AND PROBABILITY

A. ENGEL

University of Frankfurt, Frankfurt
and
Comprehensive School Mathematics Program, Carbondale

1. Introduction

Since August 1969 I have been consulting the Comprehensive
School Mathematics Program (CSMP) in Carbondale, Illinois.
In 1970-71 I have written, taught, and revised two chapters
on probability: An Introduction to probability, (125 pages)
Topics in Probability and Statistics, (75 pages). These chapters
are to be used in grades 7 and 8 and can be covered in 15
and 10 hours, respectively.

At that time I was completely ignorant of computing. In
April 1971, Burt Kaufmann, the CSMP Director, asked me to introduce computing into the program. In August-September 1971,
I have written and taught a chapter on computing: Introduction
to Computer Programming, (225 pages). It is to be used in
grade 9 and it can be covered in 25 hours.

If you teach computing you look around for suitable problems. Numerical problems are basically trivial and they re-
inforce the wrong idea that a computer is merely a super -
slide-rule. I was fortunate in that my students had some number theory and some probability. It is well known that number theory provides a wide range of excellent computer problems. For me it was a pleasant surprise to discover that
probability is even more suitable than number theory. In
fact probability is an inexhaustible source of top quality
computer problems, even for students with little or no previous knowledge of probability. One quarter of the chapter
on computing is therefore devoted to probabilistic problems.

At the high school level, computers are almost universally
misused as super-slide-rules. But a computer is really a

simulator of processes, deterministic and stochastic pro -
cesses. Simulation programs are the most instructive programs.
The computer is instructed to imitate some process, for instance to play 1000 crap games.

In junior high school, CSMP students acquire basic knowledge
in probability and computing. If this skill is not used con-
tinually it will soon be forgotten and will have no permanent
impact upon the thinking of the student. In order to prevent
this, I am writing a big volume with the title, Computer
Oriented Mathematics for grades 10-12. It will have about
30 chapters which can be read independently. Each chapter
treats some interesting problem. In studying the problem ,
students learn some computing. The computer printout leads
to conjectures. By trying to prove the conjectures, they
learn some mathematics. Since most of the problems contain
some probability the probabilistic horizon of the student
will also be extended.

The book is not meant to be covered in one block. Rather, it
is to be used parallel to the main course. Every two or three
weeks, when the student is fed up with the every day routine,
it provides an escape into the jungle of unsystematic mathematical exploration.

In this paper, I present a small selection of problems from
this book. The problems are here strongly abridged. But they
show the interaction between probability, computing and mathematics. The unabridged version of for instance problems;
 5 and 8 would take up the whole paper. In another paper
about this book, I made a totally different selection of
problems. It's title is Outline of a Problem Oriented, Computer Oriented and Applications Oriented High School Mathe -
matics Course. It is available from:
Comprehensive School Mathematics(Course)Program (610 East
College Street, University Complex, Carbondale, Illinois,
62901 USA). It is also published in International Journal of
Mathematical Education in Science and Technology, vol.4, 1973.

2. The Random Number Generator (RNG)

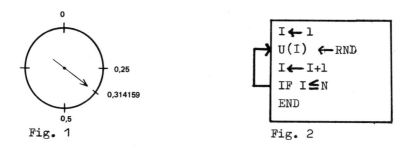

Fig. 1   Fig. 2

Every good computer has a built in random number generator.
Production of a high quality stream of random numbers is
a sophisticated art and requires extensive number theore-
tic competence. For this reason the student should think
of the RNG as the spinner in Fig. 1. At the instruction RND
the computer somehow spins this spinner and reads off a real
number between 0 and 1, accurate to 12 decimals.

The program in Fig. 2 produces and stores a stream

$$U_1, U_2, \ldots, U_N$$

of real numbers, which are uniformly distributed in the
interval (0,1), i.e. for $0 \leq a < b \leq 1$,

$$P(U_i \in (a,b)) = b-a$$

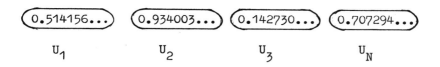

Fig. 3

These "chuncks of 100% pure chance" (Fig. 3) are the raw
material for imitating random processes. This raw material
must be processed by streching, shifting, and chopping.

Chopping is done by the function

$$\text{INTEGER PART} = [\ ] = \lfloor\ \rfloor = \text{INT}.$$

By means of the RNG we can easily simulate any discrete random device.

a) We can get the output of a fair coin by streching and chopping or by shifting and chopping. From Fig. 4 it is easy to see that each

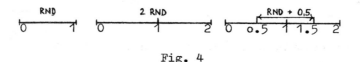

Fig. 4

of the two instructions

$$\big[2\ \text{RND}\big] \qquad \text{and} \qquad \big[\text{RND} + 0.5\big]$$

produces one spin of the fair coin in Fig. 5

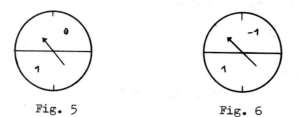

Fig. 5   Fig. 6

b) If we stretch by 2 and shift to the left by 1 we get the instructions

$$2\big[2\ \text{RND}\big] - 1 \qquad \text{and} \qquad 2\big[\text{RND} + 0.5\big] - 1,$$

which generate one spin of the spinner in Fig. 6. These are the basic instructions for imitating symmetric ran - dom walks.

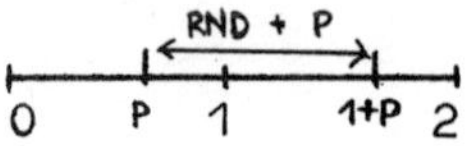

RND + P

0   P   1   1+P   2

Fig. 7

c) Fig. 7 shows that the instruction

[RND + P],

where 0 < P < 1, produces one spin of the lopsided coin
in Fig. 8.

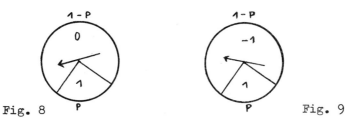

Fig. 8   Fig. 9

d) If we stretch by two and shift to the left by 1 we get
the instruction

2[RND + P] - 1

which generates one spin of the device in Fig. 9. This
instruction is used for simmulating asymmetric random
walks.

e) The instruction [6 RND] + 1 produces one roll of a
standard die.

f) The instruction [10 RND] produces a decimal random digit.

g)

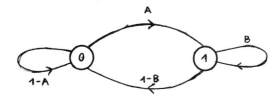

Fig. 10

Consider the two-state Markov chain in Fig. 10. The following program produces a stream of Markov dependent digits. In this case A, B and the initial state M must be fed into the computer.

```
INPUT A,B,M
U←RND
M←(1-M) [U+A]+M [U+B]
PRT M
```

Fig. 11

The instructions in a) to g) are the basic vocabulary of simulation. By using this vocabulary one can avoid branches. This makes programs fast, straightforward and easy to grasp. It is not worth remembering less frequent random devices like the ones in Fig. 12.

U ← RND

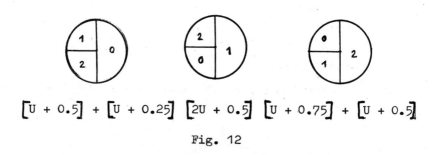

$$[U + 0.5] + [U + 0.25] \quad [2U + 0.5] \quad [U + 0.75] + [U + 0.5]$$

Fig. 12

Let us now look at some good programming problems which are at the same time good probability problems or good problems of data analysis, the data being generated by the computer.

3. Data Analysis. Equidistribution

The last n digits of the numbers in the sequence 1,2,4,8,
16,32,...... are periodic with period of lenght 4.5^{n-1}.
But what about the sequence of first digits? Is it periodic? Does each of the digits 1,2,3,.....,9 occur as first
digit? Do they occur with the same frequency?

Consider the sequence 2^n for n=0,1,2,.....,99999. Take the
first digit of each term:
1,2,4,8,1,3,6,1,2,5,1,2,4,8,1,...

Fig. 13 shows a program which counts the frequency $R(D)$
of each digit D in this sequence. Fig. 14 shows the printout produced by the program. In this table

$$H(I) = R(I)/10000$$

is the relative frequency of the digit I.

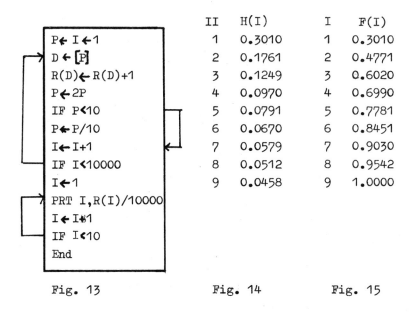

	II	H(I)	I	F(I)
P← I←1	1	0.3010	1	0.3010
D ← [P]	2	0.1761	2	0.4771
R(D)←R(D)+1	3	0.1249	3	0.6020
P←2P	4	0.0970	4	0.6990
IF P<10	5	0.0791	5	0.7781
P←P/10	6	0.0670	6	0.8451
I←I+1	7	0.0579	7	0.9030
IF I<10000	8	0.0512	8	0.9542
I←1	9	0.0458	9	1.0000
PRT I,R(I)/10000				
I ← I+1				
IF I<10				
End				

Fig. 13   Fig. 14   Fig. 15

These are the data which are to be analyzed and to be explained. They look like real life data, a limited number
of highly confusing numbers. Our situation is comparable
to that of Kepler who from the data of Tycho Brahe ex-
tracted his three laws of planetary motion.

The relative frequencies of the digits are decreasing. Is
there some regularity? We should compute

$H(n+1)/H(n)$, $H(n+1) - H(n)$, $H(1) + H(2) + \ldots + H(n)$
Let
$$F(n) = H(1) + H(2) + \ldots + H(n).$$

The table in Fig. 15 is of striking familiarity, at least
for the older generation. It is a four place table of logarithms to the base 10. Thus

$$F(n) = \log_{10}(1+n)$$
$$H(n) = F(n) - F(n-1)$$
$$H(n) = \log_{10}(1+\frac{1}{n})$$

There are many opportunities for further exploration. What
about the powers of $3,4,5,\ldots$? What about the first two
digits?

The explanation for the logarithmic law is based on the
fact that $\log_{10}a$ is irrational a for $a \neq 10^k$ and that for irrational

$$\{n\alpha\} = n\alpha - [n\alpha] = n\alpha \bmod 1$$

is equidistributed in the interval $(0,1)$.

The empirical study of equidistributions is also a good
programming problem with surprising results. Let α be irrational. The points

$$\{\alpha\}, \{2\alpha\}, \{3\alpha\}, \ldots , \{n\alpha\}$$

partition the interval $(0,1)$ in $n+1$ parts which have at
most 3 different lenghts. And the next point $\{(n+1)\alpha\}$

splits the longest of these parts. For the golden ratio φ
$= (\sqrt{5}-1)/2 = 0.6180339887$ we get the most uniformly distributed sequence $\{n\varphi\}$.   Each point splits the largest interval in the golden ratio.

To check the 3-distance-theorem the necessity for sorting
arises in a natural setting. One writes a program which computes the sequence $\{n\varphi\}$ for n=1 to 100, sorts it in in -
creasing order and prints the distances of neighboring points.
Only 3 distances will occur. The sequence should also be compared with a sequence of random numbers. Some striking differences will be found.

4. Data Analysis. Comparison of Two Distributions

Anchuria and Sikinia are separated by a big lake. The weights
of the fish in the lake are uniformly distributed between 0
and 1. Hence the instruction RND produces one fish. Each country restricts fishing by its own law.
Anchurian Fishing Law : You may go on fishing until you catch a
fish which is heavier than the preceding one. Then you must stop.
Sikinian Fishing Law: You may go on catching fish until the
total weight of your catch is greater than 1. Then you must
stop.

Let us compare these two laws. Denote by F the number of fish
a fisherman catches. We are interested in the distribution of
F. We find it empirically by studying the catches of 1000 law
abiding fishermen. The computer will do the work for us. For
Anchuria we introduce the following variables:

    I: counts the number of catches
    F: counts the number of fish in the current catch
    P: weight of the preceding fish
    N: weight of the next fish
    R(F): counts the frequency of the catch size F
    S: cumulates the F's of all 1000 fishermen

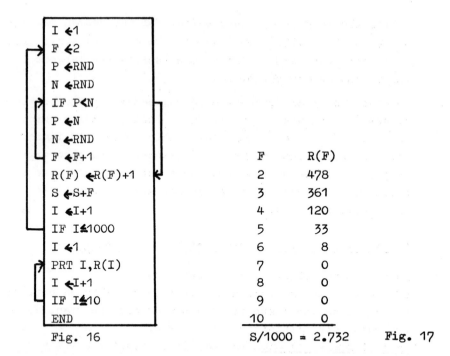

Fig. 16

F	R(F)
2	478
3	361
4	120
5	33
6	8
7	0
8	0
9	0
10	0

S/1000 = 2.732   Fig. 17

The program in Fig. 16 finds also the average of F for 1000 fishermen. Fig. 17 shows the result.

For Sikinia we introduce a new variable W, which cumulates the weights of the fish in a catch. The program for Sikinia is shown in Fig. 18 and the printout in Fig. 19.

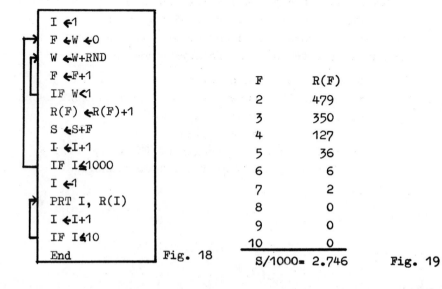

Fig. 18

F	R(F)
2	479
3	350
4	127
5	36
6	6
7	2
8	0
9	0
10	0

S/1000= 2.746   Fig. 19

If we compare the two printouts we cannot but suspect that F
has the same distribution under both laws. The small differ -
ences have the size of superimposed random poise. A skillfull
data analyst might guess the theoretical distribution from
these data. If the student takes a sample of size 10000, this
will reduce the random poise 3 times and the theoretical distribution emerges quite clearly.
Let us find the distribution of F for Anchuria. Denote by

$$U_1, \ U_2, \ U_3, \ \ldots$$

the successive weights of the fish. Obviously

$F=2$, i.e. $U_1 < U_2$ with probability $p_2 = \dfrac{1}{2!}$

$F>2$, i.e. $U_1 \geq U_2$ "   " $q_2 = \dfrac{1}{2!}$

$F>3$, i.e. $U_1 \geq U_2 \geq U_3$ "   " $q_3 = \dfrac{1}{3!}$

$F>n$, i.e. $U_1 \geq U_2 \geq \cdots \geq U_n$ " $q_n = \dfrac{1}{n!}$.

Thus the event $F=n$ occurs with probability $p_n = q_{n-1} - q_n$, which
gives   . $p_n = \dfrac{n-1}{n!}$

Fig. 20 shows the first terms of the distribution of F and
the expected frequencies $1000 \ p_n$.

n	2	3	4	5	6	7	
p_n	$\dfrac{1}{2}$	$\dfrac{1}{3}$	$\dfrac{1}{8}$	$\dfrac{1}{30}$	$\dfrac{1}{144}$	$\dfrac{1}{840}$	
$1000 p_n$	500	333	125	33	7	1	Fig. 20

Now we can find the expectation and variance of F.

$$F(F) = \sum_{n \geq 0} q_n = \sum_{n \geq 0} \frac{1}{n!}$$

$$E(F^2) = \sum_{n \geq 2} n^2 p_n = \sum_{n \geq 2} \frac{n}{(n-2)!} = \sum_{n \geq 2} \frac{n-2}{(n-2)!} + \sum_{n \geq 2} \frac{2}{(n-2)!} = 3e$$

$$E(F) = e \quad , \quad \sigma^2 = V(F) = 3e-e^2$$

$$\left[e - \frac{\sigma}{\sqrt{1000}} \, , \, e + \frac{\sigma}{\sqrt{1000}} \right] = \left[2.690, \, 2.746 \right]$$

Thus the means in Fig. 17 and 19 deviate from the expectation by not more than σ . In a similar way one can treat the case of Sikinia.

5. Simulation without the RNG.

The RNG is by no means indispensable for simulation. In fact the RNG should be avoided, if possible since it is inefficient. It wastes computation time and it gives low precision.

Let us consider a game.
Abel says to Cain: Let us toss a coin repeatedly until one of the patterns 111 or 101 occurs. If 111 occurs first you win. Otherwisw I win. What are the winning probabilities for Abel and Cain?

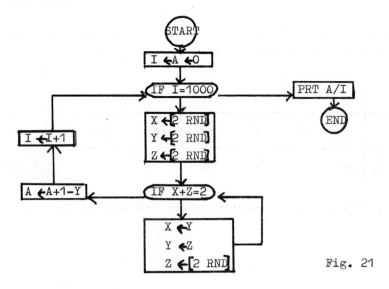

Fig. 21

We can tell the computer to play the game 1000 times and to
print the relative frequency of Abel's wins. Fig. 21 shows
the corresponding program. The computer has to perform 6800
coin tosses and·arrives at a crude estimate of Abel's winning probability.

But we can get 9 digit precision at less cost. We represent
the game by the 6-state graph in Fig. 22. Each possible game
is a random walk on this graph, starting in state O and ending in one of the states 111 or 101. Now we simulate this
linear system. Initially we place mass 1 at state O and
clear all other states. Then we pump this mass through the
system in a sequence of steps. At each step the mass at each
state is moved to the two neighboring states, each neighbor
getting the same amount. After 125 steps we print A and C.
This gives the probability that Abel and Cain have won in not
more than 125 tosses. The results are A=0.6 , C=0.4. Fig. 23
shows the simple program which performs the pumping.

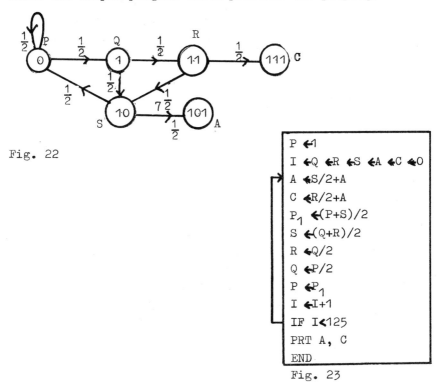

Fig. 22

```
P ←1
I ←Q ←R ←S ←A ←C ←0
A ←S/2+A
C ←R/2+A
P₁ ←(P+S)/2
S ←(Q+R)/2
R ←Q/2
Q ←P/2
P ←P₁
I ←I+1
IF I<125
PRT A, C
END
```

Fig. 23

6. Computing Expectations in Two Ways

Each time unit a coin is tossed until 1 comes up for the first time. This experiment can be represented by the graph in Fig. 24. Let E be the expected number of tosses (expected waiting time). We want the computer to find E by operating the linear system in Fig. 24.

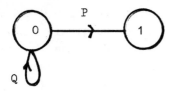

Fig. 24

Initially we place stuff of total mass A=1 in state 0. This stuff is pumped through the system in a sequence of steps one unit apart. We consider the stuff in state 0 "alive" and the stuff in state 1 "dead". If stuff of mass A has lived exactely I time units its contribution to the total life time is A I. We can find the total life time E of the initial mass 1 in two ways.

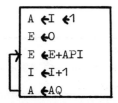

Fig. 25

a) The first method is described by the program in Fig. 25. For each I it takes the stuff that has lived exactly I time units, computes its life time and adds it to E. By tracing the program we see that the computer sums the series $E = P+2PQ+3PQ^2+4PQ^3+ \ldots$

$$A \leftarrow 1$$
$$E \leftarrow 0$$
$$E \leftarrow E+A$$
$$A \leftarrow AQ$$

Fig. 26

b) Fig. 26 describes a simpler program for finding E. At
each step all the live stuff is added to E. That is, the
computer sums the series $E = 1 + Q + Q^2 + Q^3 + \ldots\ldots\ldots$

Now suppose we have programmed some stochastic process. By
introducing into the program one additional line which
cumulates live stuff we can find expectations. For example,
take the program in Fig. 23. To get the expected duration
of the game we insert the line

$E \leftarrow E+1-A-C$

as line   3, change line   2 to

$I \leftarrow Q \leftarrow R \leftarrow S \leftarrow A \leftarrow C \leftarrow E \leftarrow 0$

and change the print statement to

PRT A, C, E

At almost no extra cost we get in this way E = 6.8 .

Let us take another example. A decimal die is rolled until each digit 0 to 9 has come up. What is the expected number of rolls to get a complete set. A simulation with the RNG
would require a complicated program and enormous computation
time. Instead we translate the problem into the graph in Fig.
27 . The states indicate the number of collected digits.

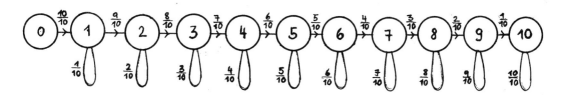

Fig. 27

Fig. 28 shows a simple program which operates the graph for
200 steps and prints E=29.2896825, exact to 9 places. The
total live stuff is also printed to check if it is 0. After
200 steps slightly less than 10^{-8} of the original stuff 1 is
still in the system. In this program roundoff errors are
quite large.

```
E ←2
X(1) ←1
I ←10
X(I) ← IX(I)+(11-I)X(I-1)
           ————————————————
                  10
I ←I-1
IF I>0
E ←E+1-X(10)
J ←J+1
IF J ≤200
PRT E, 1-X(10)
END
```

Fig. 28

7. Asymtotic Formulas

To derive asymtotic formulas advanced calculus is required
in many cases. If a computer is available asymtotic formulas
can be discovered empirically by the student. For instance,
the probability of n heads in 2n tosses of a fair coin is

$$b(n) = \binom{2n}{n} 2^{-2n} = \frac{1.3.5....(2n-1)}{2.4.6.... \quad .2n}$$

The program in Fig. 29 computes $\sqrt{n} \cdot b(n)$ for n=1,2,....,
10,20,.....,100,200,....,2000.

n	$\sqrt{n}\, b(n)$
1	0.5
2	0.53033
3	0.54127
4	0.54688
5	0.55028
6	0.55257
7	0.55421
8	0.55545
9	0.55641
10	0.55718
20	0.56067
30	0.56184
40	0.56243
50	0.56278
60	0.56302
70	0.56318
80	0.56331
90	0.56341
100	0.56348
200	0.56384
300	0.56395
400	0.56401
500	0.56405
600	0.56407
700	0.56409
800	0.56410
900	0.56411
1000	0.56412
1100	0.56413
1200	0.56413
1300	0.56414
1400	0.56414
1500	0.56414
1600	0.56415
1700	0.56415
1800	0.56415
1900	0.56415
2000	0.56415

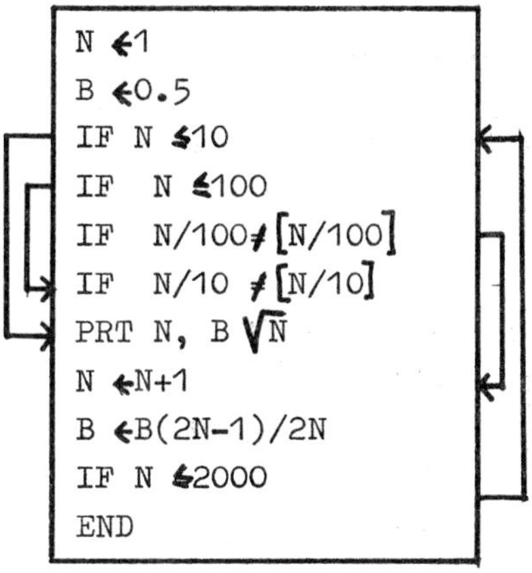

```
N ← 1
B ← 0.5
IF  N ≤ 10
IF  N ≤ 100
IF  N/100 ≠ [N/100]
IF  N/10 ≠ [N/10]
PRT N, B √N
N ← N+1
B ← B(2N-1)/2N
IF  N ≤ 2000
END
```

Fig. 29

Since $1/\sqrt{\pi} = 0.56419$ we have the asymtotic formula

$$b_n \sim 1/\sqrt{\pi n}$$

The printout shows also how good this formula is for various values of n.

From the standpoint of computer science numerical problems such as this one are not particularly instructive. The only point of interest in this program is the selective printing.

8. Generation of a Random Permutation

We would like to print the elements of the set $\{1,2,3,\ldots,N\}$ in random order so that all $N!$ permutations are equally likely. This is a nontrivial problem. We present an elegant solution. Store the numbers 1 to N in ascending order in the registers R(1) to R(N). Then a number K is chosen at random from 1 to N and R(K) is exchanged with R(N). Now the last number stays permanently in its place and the same step is applied to R(1) to R(N-1) until all numbers are assigned their permanent places in the permutation. The corresponding program is shown in Fig. 30.

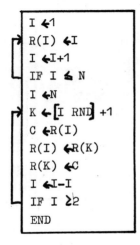

```
I ←1
R(I) ←I
I ←I+1
IF I ≤ N
I ←N
K ← [I RND] +1
C ←R(I)
R(I) ←R(K)
R(K) ←C
I ←I-I
IF I ≥2
END
```

Fig. 30

Permutations are extremely important for computer science. They lead to a great variety of interesting programs. Here are some examples:

a) A permutation is stored in R(1) to R(N). Write a program which decomposes it into cycles.
b) Find the inverse of a permutation stored in R(1) to R(N).
c) Find the order of a permutation.

d) Write a program which prints all N! permutations.

e) Write an efficient program which prints those permuta-
tions for which $|I-J| \neq |R(I) - R(J)|$.

These permutations give all possible ways of placing N non-
attacking queens on an NxN chessboard.

Another important and nontrivial problem is the selection
of a random S-sample from an N-set. Write a program which
lists all S-subsets of an N-set as N-sequences with S ones
and N-S zeros.

Many problems are permutation problems in disguise. The next
problem is a typical example.

9. Records in a Random Process

By the news media we are flooded by a never ending stream
of records, mostly record disasters: the worst flood of the
century, the most devastating storm in the past 50 years ,
the most destructive earthquake of all times, the severest
drought since 1912, and so on. Is there a trend for the
worse, or is it just what one has to expect by pure chance?

Suppose that at some location the amount of precipitation
(in inches) for the next 100 years will be X_1, X_2,...,X_{100}.
We can assume that $X_i \neq X_j$ for $i \neq j$. Let us say that a record
occurrs in the jth year if $X_i < X_j$ for all $i < j$. Suppose
that there are no systematic trends in weather, i.e. the
X_i are produced by independent spins of some spinner. The
possible number of records can be any number from 1 to 100.
But what is the expected number of records ? To study this
number we generate a sequence of 100 random numbers and
count the records in the sequence. This program is executed
100 times, the number of records in each century is printed,
as well as the average number of records. We find that in
a century there are on the average 5.17 records. Hence records are quite infrequent. (See Fig. 31)

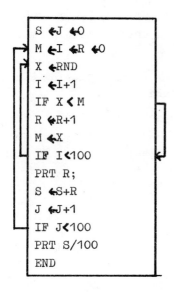

It is diffucult to guess the expected number of records by studying computer generated data. Luckily the problem has a very simple analytical solution, accessible to students with little probabilistic background. Let $U_1, U_2, \ldots, U_n$ be a sequence of random numbers. Let $R_i = 1$ if U_i is a record and $R_i = 0$ otherwise. Then

$$R = R_1 + R_2 \ldots + R_n$$

is the total number of records in the sequence. Now

$$E(R_i) = P(R_i = 1) = \frac{1}{i}.$$

Thus the expected number of records is

$$E(R) = 1 + \frac{1}{2} + \frac{1}{3} + \ldots + \frac{1}{n}.$$

For $n = 100$ we get $E(R) = 5.187$. The estimate 5.17 obtained above by simulation is excellent.

```
S ←J ←0
M ←I ←R ←0
X ←RND
I ←I+1
IF X < M
R ←R+1
M ←X
IF I<100
PRT R;
S ←S+R
J ←J+1
IF J<100
PRT S/100
END
```

```
4  7  3  8  8  6  5  7  4  4  6  8  6  5  4  4  5  2  6  5  6
8  4  8  3  2  3  6  5  4  4  5  2  5  5  2  10 7  10 2  2  9
4  5  7  6  6  5  3  8  5  3  8  3  8  4  4  4  7  2  8  3  5
5  5  4  2  8  2  4  6  8  9  5  5  6  6  10 6  5  4  3  4  6
3  8  7  6  3  5  7  4  3  2  3  7
5.17
```

Fig. 31

10. Visible Points in the Plane Lattice

The set L of all points with integral coordinates will be called the plane lattice. A point $(x,y) \in L$ will be called visible (namely visible from the origin) if its components have the greatest common divisor 1, i.e. $x \sqcap y = 1$. Let $V(n)$ denote the number of visible points in the set

$$S_n = \{(x,y) \mid 1 \leq x \leq n, \ 1 \leq y \leq n\}$$

of lattice points. Since $\# S_n = n^2$, the proportion of

visible points in S_n is $p(n) = \dfrac{V(n)}{n^2}$.

What happens to this proportion if n increases indefinitely?
Every visible point (x,y) covers infinitely many nonvisible
points (kx,ky) for k=2,3,4,.... Hence it seems plausible
that

$$\lim_{n \to \infty} p_n = 0$$

Since intuition sometimes fails we look at the visible points in
S_{20} (Fig. 32). Contrary to expectation the visible points are even-
ly spread in the square S_{20}.

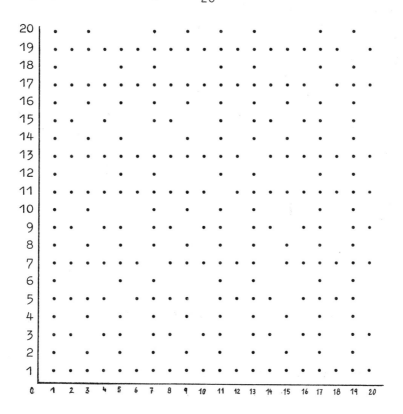

Fig. 32

Now the computer is instructed to print the table in Fig. 33. This table suggests that p_n approaches a limit for $n \to \infty$ which is approximately o.61. But we should be careful. After all we went only as far as n= 100. It is quite conceivable that far out the density of visible points decreases dramatically. To check this we consider the set S_{10000}. This time we cannot test all 10^8 points for visibility. Instead we take a random sample of 1000 points in S_{10000} and find the proportion of visible points in the sample.

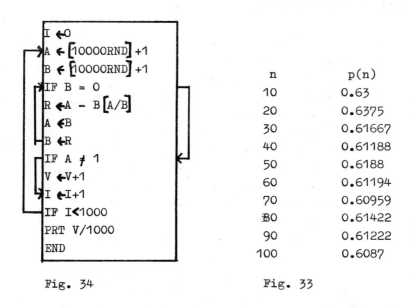

n	p(n)
10	0.63
20	0.6375
30	0.61667
40	0.61188
50	0.6188
60	0.61194
70	0.60959
80	0.61422
90	0.61222
100	0.6087

Fig. 34   Fig. 33

A run of the program in Fig. 34 resulted in the estimate

$$\hat{p}(10000) = 0.615.$$

This strongly supports our conjecture that the limit

$$\lim_{n \to \infty} p(n) = p$$

exists with $p \approx 0.61$. But we have no method of finding a precise value of p.

Let us assume that p exists. We try to find p by a plausibility argument (which can be made rigorous). The number p can be interpreted as the density of visible lattice points, or, the probability that two randomly chosen natural numbers are relatively prime.

Let S be the set of visible points. Then p is the density of S in L. Let S(d) be the set of points whose coordinates have the greatest common divisor d. We get S(d) by stretching S from the origin by the factor d. Hence S(d) has density p/d^2 in L. If $d \neq d'$ then

$$S(d) \cap S(d') = \emptyset \qquad \text{and} \qquad \bigcup_{d \geq 1} S(d) = L .$$

Hence $\quad \sum_{n \geq 1} \dfrac{p}{n^2} = 1$, or $p = 1/(\sum_{n \geq 1} 1/n^2)$.

The evaluation of this sum is a new programming challenge.

11. Infinite Expectations

Consider the symetric random walk of a particle on the line starting at x= 1. It is easy to show that with probability 1 the particle will eventually come back to 0, but the expected travel time to the origin is infinite. This is hard to understand intuitively. A computer simulation can help in understanding this counterintuitive result.

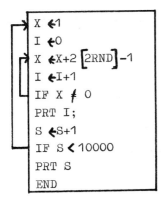

```
X ←1
I ←0
X ←X+2 [2RND] −1
I ←I+1
IF X ≠ 0
PRT I;
S ←S+1
IF S < 10000
PRT S
END
```

Fig. 35

We could start 100 random walks in 1 and count the number of steps needed to reach 0. But this may require excessive computation time. One should put a bound on the total number of steps. The program in Fig. 35 does not allow a new random walk to start beyond step number 10000. It is still dangerous to run this program. The printout in Fig. 36 is highly instructive. There are exactly 50 walks. About one half of them (26) are of length 1 as it should be. But the most remarkable observation is the tremendeous fluctuation in these numbers. The last walk which stopped the program required 14 627 steps out of a total of 15 228 steps. The average number of steps per walk is 304.56. Had we limited the step number to 10000 then the last walk would not have been completed and the average step size would be 601:49 = 12.26.

The average does not stabilize. It continues to fluctuate wildly and it tends to grow without bound as the number of walks increases.

```
3   1   1   3 1   1   1   1   1   1   1   123   1   1   15   3
37  1   3   1   3   1   1   1   1   17   3   49   5   1   3   3
1   3   205  27   1   7   1   3   1   9   1   9   23   1   1   1
14627
15228
```

Fig. 36

We need a collection of probabilistic problems satisfying the following conditions:

a) They should be interesting and important.
b) They should be analytically intractable, so that the computer is indispensable for their solution.
c) They should lead to interesting programs which are not discouragingly long, roughly not longer than 30 lines. Preferably not longer than 20 lines.

The next problem satisfies b) and c) but not a).

12. The Total Beetle Tour

A beetle starts at some vertex of a cube and performs a
random walk until it has visited all vertices. What is the
expected time for such a Total Beetle Tour ? (The beetle
covers one edge in one time unit.)

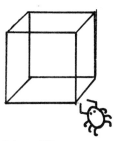

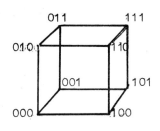

Fig. 37   Fig. 38

An analytical solution is almost hopeless since we are
dealing with a 26-state Markov Chain. Hence we simulate
1000 Total Tours with the RNG. Let us describe how we
proceed with the simulation.

To each vertex we assign a number $V=0,1,2,\ldots,7$, as shown
in Fig. 38, where the vertex numbers are written in the
binary system. The binary digits of V are stored in R(1),
R(2), R(3). We start at the vertex 000. Then at each step
a number A is chosen at random from 1,2,3, and R(A) is
replaced by 1-R(A). X(V) counts the visits to vertex = V.
C counts the total number of vertices visited so far in
the current Tour. T counts the number of steps in the
current Total Tour. J counts the number of Total Tours.
The frequency of the Total Tour length T is recorded in
Y(T). The program and the printout are shown in Fig. 39.

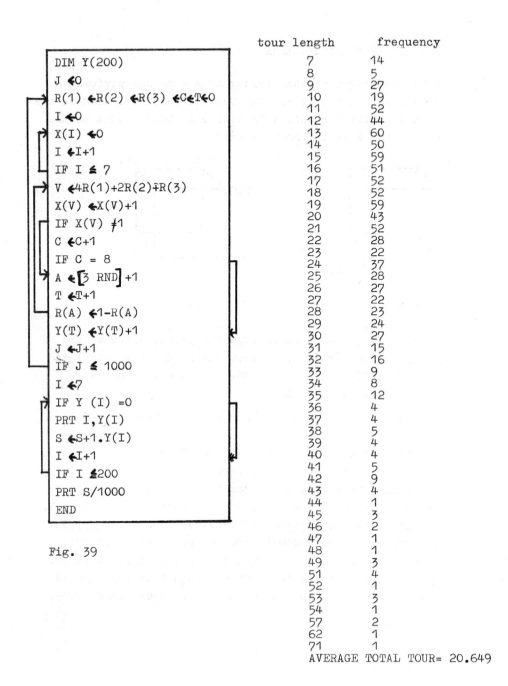

```
DIM Y(200)
J ←0
R(1) ←R(2) ←R(3) ←C←T←0
I ←0
X(I) ←0
I ←I+1
IF I ≤ 7
V ←4R(1)+2R(2)+R(3)
X(V) ←X(V)+1
IF X(V) ≠1
C ←C+1
IF C = 8
A ←[3 RND]+1
T ←T+1
R(A) ←1-R(A)
Y(T) ←Y(T)+1
J ←J+1
IF J ≤ 1000
I ←7
IF Y (I) =0
PRT I,Y(I)
S ←S+1.Y(I)
I ←I+1
IF I ≤200
PRT S/1000
END
```

Fig. 39

tour length	frequency
7	14
8	5
9	27
10	19
11	52
12	44
13	60
14	50
15	59
16	51
17	52
18	52
19	59
20	43
21	52
22	28
23	22
24	37
25	28
26	27
27	22
28	23
29	24
30	27
31	15
32	16
33	9
34	8
35	12
36	4
37	4
38	5
39	4
40	4
41	5
42	9
43	4
44	1
45	3
46	2
47	1
48	1
49	3
51	4
52	1
53	3
54	1
57	2
62	1
71	1

AVERAGE TOTAL TOUR= 20.649

TEACHING OF STATISTICS AT THE SECONDARY LEVEL - AN INTERDISCIPLINARY APPROACH

C. RADHAKRISHNA RAO

Indian Statistical Institute, Calcutta

1. Introduction

In her book on Studies in the History of the Statistical Method, published in 1931, Helen M. Walker referred to education in statistics as follows:

> Educational statistics is the offspring of a varied ancestry. The greed of ancient kings enumerating their people for taxation; the panic of an English sovereign during London plague; the cupidity of professional gamblers; the scientific ardor of psycho-physicists; the labour of mathematicians and astronomers and physicists and actuaries; the enthusiasm of the students of social phenomena; the disciplined imagination of the biologists; and the vision of the educators planning a new science of education; from these has educational statistics descended.

This provides a succinct summary of the different aspects of statistics, some of which received parallel development and which together provided the broad base for the evolution of statistics as a separate discipline. Walker also referred to the general awakening in the importance of statistics and the wide scale on which courses in statistics have been introduced in the universities all over the world.

Forty years later William H. Shaw, in his Presidential Address to the American Statistical Association, entitled Paradoxes, Problems and Progress, examined the current position of statistics from three points of view - used in the plural as data, used in the singular as statistical methodology, and used in the original etymological sense as Political Science. He pointed out some of the paradoxes and problems in the use of statistics, expressed his concern

about the future progress of the subject, and concluded
by charging the statisticians with some responsibilities
if "statistics,however defined, is - or are - to remain a
truly variable and socially meaningful science".

We have to examine the future education in statistics and
training of statisticians keeping in view the varied
ancestry of statistics referred to by Walker and the hopes
and fears raised by Shaw about the future progress of statistics.

It is unfortunate that, at present, statistical education
at higher levels in the universities is in a most chaotic
condition. Some universities have a separate department
for statistics, while others do not, indicating that there
is some difference of opinion about the status of statistics
as a separate discipline. The situation is further confused.
Within the same universities, there are economic statisticians teaching statistics to students of economics, psychological statisticians teaching statistics to students of
psychology and so on. Each department insists that statistics
should be taught to their students in the language of the
discipline in which they are specialising. Besides these
different castes of statisticians there is the Brahmin
called the mathematical statistician located in the mathematics or statistics department, who claims to teach what
he calls statistical methods in the abstract applicable to
all disciplines although he may have no experience of how
exactly these methods are used in any practical problem.
There is very little communication between these castes
within the same university. Each department formulates its
own syllabus in statistics and uses text books specifically
designed to cover the recommended topics. In such a situation, statistical courses in any department would naturally
be biased towards statistical techniques currently fashion-
able in research in a particular discipline. This is far
from satisfactory as such training would considerably re-
strict the scope of enquiry in any particular problem and

thus effect the progress of research both in statistics and the concerned discipline.

Heavy emphasis on factor analysis in the treatment of psychological data neglecting other useful multivariate techniques and on model building in economics, utilizing economic variables on which information exists instead of exploring the possibilities of acquiring fresh and new types of data throwing light on economic phenomena, are some of the consequences of the narrow and compartmentalized training in statistics.

2. Statistics and its role in research

The situation must be remedied. In seeking for a solution let us ask ourselves as to what constitutes the subject matter of statistics as a separate discipline and how it is related to the sciences and technology. It is common knowledge that statistics has contributed immensely to progress in scientific and technological research. In some scientific investigations statistics is still the only available approach. Statistics has helped immensely in increasing industrial productivity by an optimum use of available resources and individual efforts. It is useful in regulating our own individual lives properly, and that of the society in which we live. The horizons of application are expanding rapidly contributing at every stage to the welfare of mankind. (See Mosteller and others, 1971).

The subject matter of statistics may be broadly defined as information gathering and information processing. As such it is basic to all scientific inference for advancement of natural knowledge and decision making in technology and practical life. To the pioneers like R. A. Fisher, Karl Pearson, and P. C. Mahalanobis, statistics is a key technology which has a fundamental role to play in any activity involving the logical cycle - choice of a hypothesis, collection of facts, drawing inference and making a new hypothesis.

Statistics ceases to have a meaning if it is not related to
any practical problem. There is nothing like a purely statistical problem which statistics purports to solve. The
subject in which a decision is taken is not statistics. It
is botany or ecology or geology and so on. Statistics can
flourish only in an atmosphere where it is in demand and it
can develop only in trying to solve problems in other disciplines. Equally, good scientific research and optimum decision making are possible only when facts are gathered in
an organized way with objectives clearly defined, data are
summarized without entailing loss of information and interpreted in a logical way, all of which come under the realm
of statistics. What is of paramount importance is then the
interface between statistics and scientific disciplines, or
developing proper communication between statisticians on
one hand and users of statistics on the other ? How is this
going to be achieved? It is in this connexion the education
in statistics at secondary levels receives importance.

It is not uncommon that a statistician analysing data
collected by a scientist tends to formulate questions and
hypothesis to suit the statistical techniques with which
he is familiar, which the scientist may find irrelevant to
his problem under investigation. The scientist is often im-
patient, wants to draw conclusions from insufficient evidence
and is generally unable to pose questions in such a way that
they are readily amenable to statistical tests.

Successful collaboration between a statistician and a scientist is possible only if each understands the language of
the other. The statistician should try to incorporate any
special theoretical knowledge peculiar to the problem in the
statistical analysis of data. Equally, the scientist must be
aware of the statistical methods to guide him in the collection of data and to provide the statistician with the ne -
cessary information for using appropriate statistical tools.

I have mentioned about communication difficulties between
scientists and statisticians and the need to overcome them.
The same is true in other areas such as industry, administration and management, where statistics is being routinely
used for increasing productivity, taking decisions, and
improving efficiency. The technicians, administrators and
executives should have sufficient knowledge to understand
the statistical summaries (graphs, charts, averages, indices etc.) provided by statisticians and take appropriate actions based on them. Of course, the statistician must acquire enough knowledge of the subject area to collect relevant data and provide proper statistical advice.

3. Need for statistical education at the secondary level.

If we recognize the need for effective communication between
the users of statistics and statisticians, then we have to
seek for a solution in the proper education and training of
both these groups. The users should be exposed to the fundamental concepts and methods of statistics emphasizing its
role in research and in taking decisions, while specialising
in their respective subjects. The statisticians should acquire
good knowledge of some basic science and the nature of re -
search investigations in order to understand the effective
role played by statistics in acquiring new knowledge.

Describing statistics as the key technology of our century
the late Sir Ronald Fisher made the following observations
on the training and education of statisticians.

> The technologiest must talk the language of both the
> scientist and the technician. His education must be
> broader than theirs, though at points less intensive.
> He has to see both sides of the fence, and is the
> channel through which alone the skills of the others
> can be made more effective.

> Teaching, instruction, or training in statistics, at
> whatever level is bound to gravitate to an exhibitio-
> nism in useless mathematics, unless it is linked as
> intimately as may be, on the one side with the fact -
> finding projects in the traditional fields of demo -

graphy and economics, and on the other side with opportunities to gain first hand familiarity with at least
some field in the natural sciences. Moreover, the
science with which the student is to become acquainted
must be genuine research in its own right, not what is
eloquently called a <u>mock up</u> for the use of students only.

To sum up, in my opinion, teaching of statistics at the secondary level will go a long way in remedying the present
situation.

a) It would impart a proper understanding of statistical
methods which would be of value to those students who
subsequently pursue research in the natural and social
sciences or who enter the field of commerce and industry.

b) It would provide a good foundation of statistical know -
ledge and its usefulness in science and technology to
those who wish to have further education and training
with a view to becoming professional statisticians.

c) It would also help the non-specialist in acquiring a broad
appreciation of statistics and statistical methods, which
would give him a deeper understanding of social and economic problems and help him in the making of policy de -
cisions.

d) Early exposure to statistical concepts and inductive inference would help in developing the capacity for critical thinking and creating interest in the pursuit of
basic problems.

Intellectual values as well as practical usefulness, the two
traditional reasons which usually decide the choice of academic subjects, are both in favour of exposing students to
statistics at an early stage of their education before they
begin to specialise in any subject.

4. <u>Syllabus for statistics at the secondary level</u>

What should be the content of a course in statistics at the
secondary level? What is the best way of presenting statistical concepts and methods? To what extent can we integrate the

teaching of statistics with other subjects?

With reference to the first question, I would suggest the following as possible topics for courses in statistics at the secondary level:

(1) Levels of measurement
(2) Theory of errors
(3) Data collection and scrutiny
(4) Graphical techniques
(5) Discrete probability models
(6) Sample surveys
(7) Elements of design of experiments
(8) Elements of statistical inference.

I have chosen these topics for a number of reasons:

1. They have an important part to play in any research investigation involving measurements, collection of data in an optimum way subject to available resources and summarisation of data to reveal relevant facts.

2. The logical concepts behind these topics are simple and will appeal to the young minds in the formative stages of their thinking processes.

3. They can be taught - and this is important - as an integral part of courses in any discipline.

I would like to provide brief comments on each topic and indicate how they may be presented to students. Some of the topics are of basic importance and could be taught to all students. Others are somewhat specialized in character and would be useful to those who eventually choose a research career in statistics or some branch of the sciences.

4.1. Levels of measurement

It is not generally understood that all measurements need not to be of the quantitative type such as length, volume, weight etc.. The purpose of measurement is to identify or classify an object or compare one object with another.

Measurements are generally classified into four distinct
scales; nominal, ordinal, interval, and ratio. It would be
of interest to discuss by examples various situations in
which these four different types of measurements are applicable and explain the nature of information they supply. For
instance, if I describe my friend as a Dutchman, Associate
Professor and five feet tall, the three attributes are indeed three measurements taken on an individual, the first
one being on the nominal, the second on the ordinal (in the
hierarchy of academic positions) and the third on the ratio
scale.

4.2. Theory of errors

The need to study theory of errors was admirably stated in
a report by "Student" (the late W.S. Gosset) submitted to
the Board of Directors of the Guiness Brewery. The document,
dated 3 November 1904, opens with the following paragraph:

> The following report has been made in response to an in-
> creasing necessity to set an exact value on the results
> of our experiments, many of which lead to results which
> are probable but not certain. It is hoped that what
> follows may do something to help us in estimating the
> Degree of Probability of many of our results, and enable
> us to form a judgment of the number and nature of the
> fresh experiments necessary to establish or disprove
> hypotheses which we are now entertaining.

> When a quantity is measured with all possible precision
> many times in succession, the figures expressing the
> results do not absolutely agree, and even when the aver-
> age of results, which differ but little, is taken, we
> have no means of knowing that we have obtained an actually
> true result, the limits of our powers are that we can
> place greater odds in our favour that the results ob -
> tained do not differ more than a certain amount from the
> truth.

> Results are only valuable when the amount by which they
> probably differ from the truth is so small as to be in-
> significant for the purposes of the experiment.What the
> odds should be depends:

> (1) On the degree of accuracy which the nature of the
>   experiment allows, and
> (2) On the importance of the issues at stake. It may

seem strange that reasoning of this nature has not
been more widely made use of, but this is due:

a) To the popular dread of mathematical reasoning.
b) To the fact that most methods employed in a Lab -
    oratory are capable of such refinement that the re-
    sults are well within the accuracy required.

Unfortunately, when working on the large scale, the interests are so great, that more accuracy is required, and, in
our particular case, the methods are not always capable of
refinement. Hence the necessity of taking a number of inex-
act determinations and of calculating probabilities.

Differences that arise in parallel (repeated) measurements
on the same object (errors of measurement) and in measurements taken by different observers, or by the same observer using different instruments on the same object (observer or instrument bias) must be distinguished from differences that may be observed between measurements on different objects (natural variation between objects). Detection
and estimation of differences between objects in the pre -
sence of errors of measurements and possible bias is of
prime importance in any investigation.

A very simple experiment can be conducted to demonstrate all
these aspects of measurements and to draw valid inferences
in any field of study. I would like to describe one such experiment conducted to examine whether the hight of an individual (stature) decreases during the day.

Table 1 gives the measurement (in m.m) taken on 41 students
by two anthropologists (designated as B and M) trained by
two different experts in measuring the stature of an indi -
vidual defined in a particular way. Each anthropologist
measured each student twice in the morning when he woke up
and twice before he went to bed. The second set of measurements were taken after completing the first set on all the
students, to ensure independence of repeated measurements.
Table 2 gives the errors in repeated measurements, bias between anthropologists and the time effect (between morning

and evening) in each case. This simple analysis brings out
clearly the observer bias, the measurement by M being con-
sistently larger than that by B with an average difference
of 2.7 m.m., and the effect of time, which is consistent
between observers and which is of the order of 9.6.m.m. on
the average.

It appears that we are about 1 c.m. longer in the morning
than in the evening.

Table 1 Morning and evening stature (in m.m.) of students

st. no.	investigator B evening e		morning m		investigator M evening e		morning m		average stature	average m - e
1.	1717	1723	1727	1728	1724	1717	1730	1730	1724.5	8.50
2.	1528	1531	1535	1538	1526	1529	1540	1540	1533.4	9.75
3.	1454	1451	1462	1462	1454	1451	1463	1462	1457.4.	9.75
4.	1778	1775	1783	1780	1775	1778	1783	1784	1779.5	6.00
5.	1664	1663	1671	1669	1671	1673	1672	1672	1669.0	4.00
6.	1573	1569	1580	1581	1572	1570	1583	1583	1576.4	10.75
7.	1662	1664	1673	1672	1667	1665	1676	1674	1669.1	9.25
8.	1709	1705	1722	1718	1709	1711	1723	1724	1715.1	13.25
9.	1633	1633	1648	1647	1639	1639	1646	1645	1641.2	10.50
10.	1782	1780	1796	1793	1782	1783	1793	1793	1787.8	12.00
11.	1815	1816	1823	1825	1814	1811	1826	1827	1819.6	11.26
12.	1788	1783	1804	1801	1788	1789	1800	1801	1794.2	14.50
13.	1729	1727	1739	1741	1732	1731	1746	1744	1736.1	12.75
14.	1711	1712	1722	1720	1713	1710	1721	1720	1716.1	9.25
15.	1714	1713	1725	1729	1723	1719	1728	1731	1722.8	11.00
16.	1743	1740	1752	1754	1744	1744	1753	1756	1748.2	11.00
17.	1715	1714	1722	1725	1718	1720	1728	1727	1721.1	8.75
18.	1593	1589	1598	1596	1593	1592	1596	1602	1595.1	5.75
19.	1747	1744	1755	1753	1748	1749	1758	1759	1751.6	9.25
20.	1663	1669	1669	1670	1662	1665	1679	1678	1668.2	11.50
21.	1676	1675	1688	1687	1678	1679	1689	1691	1682.9	11.75
22.	1678	1678	1687	1683	1686	1682	1691	1692	1684.9	7.75
23.	1610	1610	1620	1617	1617	1617	1629	1626	1617.1	7.25
24.	1665	1663	1679	1678	1669	1671	1679	1680	1673.6	10.73
25.	1552	1549	1554	1555	1551	1549	1560	1560	1553.8	7.00
26.	1694	1692	1702	1703	1699	1701	1706	1708	1700.5	8.60
27.	1619	1615	1631	1632	1621	1621	1634	1634	1625.9	13.75
28.	1583	1581	1586	1587	1579	1583	1588	1587	1584.2	5.50
29.	1587	1591	1596	1593	1593	1591	1600	1601	1594.6	8.25
30.	1583	1582	1591	1589	1585	1584	1596	1593	1587.9	8.75
31.	1709	1708	1717	1716	1710	1710	1722	1723	1714.4	10.25
32.	1792	1794	1803	1804	1797	1797	1812	1811	1801.2	12.50
33.	1619	1618	1622	1624	1621	1620	1624	1626	1621.8	4.50
34.	1692	1694	1701	1705	1695	1697	1708	1707	1699.9	10.75
35.	1687	1688	1694	1691	1683	1680	1693	1693	1689.5	6.75
36.	1779	1783	1790	1794	1785	1781	1798	1799	1788.6	13.25
37.	1623	1626	1642	1640	1629	1632	1646	1646	1636.1	14.75
38.	1672	1669	1674	1673	1665	1667	1679	1683	1672.8	9.00
39.	1637	1638	1649	1645	1645	1646	1648	1648	1644.6	6.25
40.	1609	1607	1618	1619	1609	1608	1623	1620	1614.1	11.75
41.	1720	1721	1728	1728	1722	1722	1728	1728	1724.4	6.25

Table 2 Bias and time effect

	Errors of measurement				Bias		Time effect	
	B		M		M-B		B	M
St.no.	e_1-e_2	m_1-m_2	e_1-e_2	m_1-m_2	e - e	m - m	m - e	m - e
1	-6	-1	7	0	0,5	2,5	7,5	9.5
2	-3	-3	-3	0	-2.0	3.5	7.0	12.6
3	3	0	3	1	0.0	0.5	9.5	10.0
4	3	3	-3	-1	0.0	2.0	5.0	7.0
5	1	2	-5	0	7.0	2.0	6.6	1.5
6	4	-1	2	0	0.0	2.5	9.5	12.0
7	-2	1	2	2	3.0	2.5	9.5	9.0
8	4	4	-2	-1	3.0	3.5	13.0	13.5
9	0	1	0	1	6.0	-2.0	14.5	6.5
10	2	3	-1	0	1.5	-1.5	13.5	10.5
11	-1	-2	3	-1	-3.0	2.5	8.5	14.0
12	5	3	-1	-1	3.0	-2.0	17.0	12.0
13	2	-2	1	2	3.5	5.0	11.0	13.5
14	-1	2	3	1	0.0	-0.5	9.5	9.0
15	1	-4	4	-3	7.5	2.5	13.5	8.5
16	3	-2	0	-3	2.5	3.5	11.5	10.5
17	1	-3	-2	1	4.5	4.0	9.0	8.5
18	4	2	3	-6	2.6	2.0	6.0	5.5
19	3	2	-1	-1	3.0	4.5	8.5	10.0
20	3	-1	-3	1	2.0	9.0	8.0	15.0
21	1	1	-1	-2	3.0	2.5	12.0	11.5
22	0	2	-2	-1	6.5	5.5	8.0	7.5
23	0	3	0	-6	7.0	4.5	8.5	6.0
24	-3	1	-2	-1	3.5	1.0	12.0	9.5
25	3	-1	2	0	-0.5	5.5	4.0	10.0
26	2	0	-2	-2	7.0	5.0	9.0	7.0
27	4	-1	0	0	4.0	2.5	14.5	13.0
28	2	-1	-4	1	-0.5	1.0	4.5	6.5
29	-4	2	2	-1	3.0	3.5	8.0	8.5
30	1	2	1	3	2.0	4.5	7.5	10.0
31	1	1	0	-1	1.5	6.0	8.0	12.5
32	-2	-1	0	1	4.0	8.0	10.5	14.5
33	1	-2	1	-2	2.0	2.0	4.5	4.5
34	-2	-4	-2	1	3.0	4.5	10.0	11.5
35	-1	3	-3	0	-3.0	0.5	5.0	8.5
36	-4	-4	4	-1	2.0	6.5	11.0	15.5
37	2	2	-3	0	3.5	5.0	14.0	15.5
38	3	1	-2	-4	-4.5	7.5	3.0	15.0
39	-1	4	-1	1	8.0	1.5	9.5	3.0
40	2	-1	1	3	0.5	3.0	10.5	13.0
41	1	2	0	0	1.5	1.0	6.5	6.0
Average	0.88	0.32	-0.10	-0.46	2.39	3.10	9.23	9.94

4.3 Scrutiny of data

Statisticians are very often required to deal with data collected by a scientist, which may be subject to gross recording errors specially when observations are taken under difficult conditions as in a field investigation and when measurements cannot be repeated. In such cases, application of statistical techniques might give misleading or wrong results. It is, therefore, necessary to examine and make sure that the data supplied are genuine and free from recording errors. A dialogue with the concerned scientist and a reference to original record books might be of help in some cases. There are no routine statistical checks for scrutiny of data. Each body of data may require special techniques and much depends on the experience and skill of the statistician. It would be useful to provide such an experience at a very early stage of a statistician's training. Some useful techniques in scrutinizing original data are discussed by Fisher (1936), Mahalanobis (1933), Mahalanobis, Majumdar and Rao (1949), Mukherji, Rao and Trevor (1955), Mazumdar and Rao (1958) and Rao (1972).

A statistician should also receive training in the preparation of a schedule (or a questionnaire or a data sheet) for recording experimental or survey data. The importance of introducing cross checks in schedules for examining internal consistency of data should be emphasized. Care must be taken to see that a schedule carries **all the necessary information,** such as identification of the subject (or individual) examined, units of different measurements, identification of the observer, time and place where the subject is observed and so on, which will be of help in scrutinizing the primary record as well as in the proper interpretation of the statistical analysis of data.

Training in the preparation of schedules and scrutiny of data are possible only when the students are directly involved in generating data from a real experiment in some branch of science. The experience, skill and knowledge of the type we

are discussing cannot be imparted through merely class room
lectures using examples from published reports. Such examples
are, however, useful in explaining different aspects of scrutiny to the students and stressing their importance.

4.4 Graphical techniques

The book on <u>Statistical Methods for Research Workers</u> by the
late Sir Ronald Fisher has a remarkable opening chapter entitled, Diagrams, which I think has not received due recognition by the statisticians. Other serious books on statistics
do not mention diagrams and graphs. Graphical presentation
of data must be regarded as an integral part of data analysis
and I hope future books on statistics give more attention to
it. Recently there has been some awareness of its usefulness
and a number of papers are being written on the subject. For
a survey of recent work on graphical techniques, see Gnanadesikan (1973).

Besides being of value "in suggesting statistical tests and
in explaining the conclusion based on them", (as mentioned
by Fisher), graphical presentation of data provides an in -
sight into the problem under investigation and sometimes enable us to draw conclusions without or with a little further
analysis. A graph provides a better summary than averages
and indices and is, therefore, a valuable tool for preliminary examination of data leading to the choice of an appropriate model and possible transformation of the variables
to simplify the statistical analysis. A graph brings out the
outliers clearly and is thus a useful device in scrutiny of
data and in looking for <u>contamination</u> in data.

Graphical methods have universal applicability and they can
be taught without any advance preparation of students in statistical methodology or mathematics. For example, the concept
of a control chart could be introduced as a graphical techniques and its usefulness in industrial production explained.

4. 5 Discrete probability models

I have discussed, elsewhere (chapter 13 in Råde, 1970), how
discrete probability models can be introduced in a natural
way while investigating random phenomena. Data may be generated by drawing cards, throwing dice etc., and the concept
of probability could be developed in relation to these random experiments. The relevance of these studies to practical
problems must be emphasized. In the publication referred to
above I showed how a comparison of coin tossing data with a
sequence of male and female births observed in a maternity
hospital would enable one to draw an inference on the mechanism
of sex determination.

A number of discrete distributions arise in biological studies
which are interesting and which may lead to some fundamental
discoveries. For instance the petal numbers of flowers borne
on the same tree may show variation. In a recent study by
Roy (1959), frequency distributions of the petal number have
been obtained for flowers from the same tree at different
points of time over a period of one year. Mean and variance
of petal number were computed from each frequency distribution
and plotted against the corresponding time point. Both the
mean and variance showed seasonal variation raising a number
of interesting questions.

4.6 Sample surveys

Survey sampling offers an excellent scope for training a statistician in various aspects of statistical work. A sample
survey involves:

1) Preparation of schedule
2) Drafting of manual, giving the concepts and definitions
   involved, and instructions to investigators in the
   choice of sampling units and recording observations
3) Training of investigators

4) A pilot survey to test draft schedules, to provide field experience to investigators and to obtain information for the preparation of the budget for a survey

5) Choice of frame

6) Design of survey

7) Planning of field investigation

8) Supervision of field work and cross checks

9) Scrutiny of primary data and tabulation

10) Reporting of survey results.

Courses on sample surveys generally concentrate only on the Design aspect and do not refer to the other important practical aspects in which the statisticians must acquire expertise.

The best way of teaching survey sampling is to choose some problem and let the students conduct an actual survey. The general survey methodology can be developed while planning the survey and reasons may be given for taking particular decisions at each stage. Stress should be laid on the practical aspects of a sample survey.

4.7 Design of experiments

Data collected by scientists are often found to be defective due to lack of controls, insufficient replication and non - random assignment of treatments to experimental units. So , it would be extremely useful to include elements of design of experiments in a compulsory course. The basic principles of design of experiments can be introduced through suitable examples like Fisher's tea-tasting experiment and diurnal variation in human stature described in section 4 of this paper. The concepts of local control, replication, randomization and concomitant observations may be introduced with reference to concrete examples and their importance in ex -

perimental investigations explained.

It would also be useful to discuss factorial experiments
and the importance of study of interaction between factors.

4.8 Statistical inference

Discussion of sample surveys and design of experiments needs
some knowledge of statistical inference like point and interval estimation and testing of hypothesis. For this purpose
it may not be necessary to consider general theories of inference. But the inductive aspect of statistical inference,
or making statements under uncertainty must be emphasized.

Properties of test criteria with respect to power function
(for alternative hypotheses), and of estimators with respect
to bias, consistency and variance can be demonstrated by suitably simulated experiments using a computer. Formal mathematical treatment of these topics is neither necessary nor desirable.

5. Approach to the teaching of statistics

While advocating introduction of courses in statistics at the
secondary level, I must emphasize the interdisciplinary approach in teaching the subject. In my opinion statistics should
not be taught as a separate discipline, as the sole purpose
is to inculcate in the students quantitative approach and
thinking, and this cannot be done without reference to real
problems. We have to develop special methods of teaching ,
which provide opportunities to impart knowledge of statistical methods while guiding a student to think through a real
problem in some discipline. Early experience in determining
what information is relevant to solve a given problem, collection of data which contain the desired information, critical
review of data and processing of information will be of great
value to an individual whether he becomes a scientist or a
technologist or an executive. For this purpose laboratory and
field work should be made an integral part of instruction.

The task is not an easy one. The material for instruction
through lectures, demonstration and field (laboratory) experiments should be problem oriented. I have given at least
one example of an experiment in anthropometry, which can
form the basis of discussion for a variety of statistical
concepts and methods while leading to new knowledge about
the diurnal variation in human stature. It is also an experiment which raises a number of other problems for in -
vestigation, which is characteristic of true research. For
instance, if the human body stretches by 1 c.m. while at
rest it might be of interest to know, which portion of the
body elongates most. In my earlier paper Rao (1970), I have
considered another experiment which can provide a proper
introduction to discrete distributions, study of random
phenomena and elements of inference, while at the same time
throwing information the mechanism of sex determination. It
should be possible to illustrate a wide variety of statistical
methods through a few selected experiments and field projects
in some branch of science.

The method of teaching I am advocating should not be con -
fused with what is called training in statistics or pro -
viding work experience. This is neither desirable nor feasible
at the secondary level of education. The object is to motivate the student to "stretch his mind", develop a spirit
of enquiry, understand how new knowledge is acquired and cul-
tivate some degree of precision in thinking.

I have mentioned a number of topics in statistics which
could be taught to students at the secondary level. How -
ever, it is not suggested that statistics should be taught
as a separate subject with these topics as contents of the
course. Some of the topics could form part of syllabus in
economics, some in biology, some in mathematics and so on.
How exactly such infiltration of statistical concepts can
be achieved needs careful discussion and should be examined
jointly by competent scientists and statisticians.

References

(1) Fisher, R.A. (1936): Has Mendel's work been rediscovered? Annals of Science, 1, 115-137.

(2) Gnanadesikan, R. (1973): Graphical methods for informal inference in multivariate data analysis. Proc. 39th Session of International Statistical Conference.

(3) Mahalanobis, P.C. (1933): A revision of Risley's anthropometric data relating to the tribes and castes of Bengal. Sankhya, 1, 76-105.

(4) Mahalanobis. P.C., Majumdar, D.N. and Rao, C.R. (1949): An anthropometric survey of the United Provinces, 1941 - A statistical study. Sankhya, 9, 90-324.

(5) Majumdar, D.N. and Rao, C.R. (1958): Bengal anthropometric survey, 1945 - A statistical study, Sankhya, 19, 201-408.

(6) Mosteller, F., and others (1972): Statistics: A guide to the Unknown. Holden - Day, Inc.,U.S.A.

(7) Mukherji, R.K., Rao,C.R. and Trevor, J. (1955): The Ancient Inhabitants of Jebal Moya, Cambridge University Press.

(8) Råde, Lennart (1970): The Teaching of Probability and Statistics. Almquist and Wicksells Boktryckeri AB.

(9) Rao, C.R. (1970): A multidisciplinary approach for teaching statistics and probability. Chapter 13, pp. 251-272 in The Teaching of Probability and Statistics. Edited by L.Råde.

(10) Rao, C.R. (1972): Data, Analysis, and Statistical thinking. Chapter 17 (pp. 383-92) Economic and Social Development. Vora and Co.

(11) Roy, S.K. (1959): Regulation of Morphogenesis in an Oleaceous Tree, Nyctanthes arbor - tristis, Nature, Vol. 183, pp. 1410 - 1411, May 16.

DISCUSSION

H. Aigner: In countries where a large percentage of the age-
bracket is in secondary school and many graduates will go
into non-scientific jobs, students' intellectual capacity will
not permit a rigorous treatment of statistical concepts in a
course compulsory for all. It is therefore suggested that the
common core of statistics taught to all should limit itself
to showing that it is indeed useful to treat matters that are
neither absolutely and demonstrably true nor absolutely and
demonstrably false by quantitative methods. Students whose
appetite for statistics appears to be wetted by such an introductory approach might be offered the more advanced course,
which might indeed be made compulsory for those wishing to
study certain disciplines at university level.

L. Schmetterer: It is obvious that Dr. Aigner has a lot of
experience, but I still support the ideas of Professor Rao,
namely to introduce a basic compulsory course in statistics
at the secondary school level. Statistics is today a very important subject which should in my opinion not only be taught
occasionally hidden in the teaching of other subjects, but
also in its own right. Only in this way, does it seem possible
to me to make young students acquainted with the basic facts
of statistics. It may be that Professor Rao's programme is a
little too ambitious. Moreover, I think such a basic course
in statistics should start with an explanation of the different levels of measurement which occur in connection with statistical problems. This suggestion comes from my experience
which I aquired when teaching statistics at a very elementary
level at Austrian summer schools.

B. Benjamin: It is very easy to organise simple sample surveys. Students can investigate, for example, the wishes of
other students with regard to meals in the refectory or their
modes of travel to school. The students can draw up questionnaires, take samples, carry out interviews and make simple
analysis of results. There is also the possibility of demonstrating estimation by asking students to estimate the total
number of words (or words beginning with a particular letter)
in a text by taking sample lines.

A MOBILE MINI-COMPUTER FOR SCHOOLS COMPUTING

R. A. S. WHITFIELD

Huntingdon Research Centre, Huntingdon

It is not easy to separate statistics from mathematics so far as the use of the computer is concerned; and since both subjects would be studied on the computer were one avail - able to a school it would, I think, be a mistake to attempt such a separation. In what follows I take it for granted that, if possible, some training in the use of computers should be available in schools and that statistics would form a natural part of that training. I therefore propose to review some of the current trends in computer education, to try to assess the aims of school computing, and finally to examine some of the techniques available, and in particular to propose a slightly different approach based on a mobile mini-computer.

We have had in the U.K. the benefit of a government-spon - sored committee whose brief was entitled "Computers and the Schools". The report (3) in fact only referred to Scotland, but is generally regarded as a useful guide-line for the whole country. There is an interim and a final report and they are known together as the "Bellis Report" after the name of the Committee Chairman. The interim report sub - divided the subject into three headings:

1) Teaching about computers and computing;
2) The use of the computer as an aid to teaching (CAI, CBL, etc.);
3) The use of computers in school administration.

This list may appear rather incongruous in that it links the teaching curriculum with the school organization, thanks to the two common key words: school, computer. In fact the

Committee referred the second item to another government
body, the National Council for Educational Technology, and
never reported at all on the third item. This appears to be
an example of the confusion which sometimes occurs in com -
puting matters. I make this point because I think it is
highly relevant in forming computing policies, in parti -
cular an educational policy.

Even the loose general heading "computer education" embraces
such disparate topics as:

1) The Leeds project for teaching English to immigrant children using a medium-size computer, together with audio -
   visual techniques (1)
2) Studying mathematical functions with the help of a computer.
3) Teaching the rudiments of a computer by means of a
   "matchbox computer" (2).

In the first example researcher and pupil are both using the
computer as a tool. We may learn more about how children learn
English, and the children may learn more English. What opinions
and attitudes the children form towards computing is not be -
ing studied. The third example is at the other extreme: learning a complicated subject with simple equipment. As such it
is likely only to be taught and to appeal to the more advanced stream. With the right approach to computing in schools
I believe these two examples would have a more natural relationship.

Formulating the aim of computer education is a delicate matter,
depending very much on looking ahead successfully. Mr. F. J.
Stockwell, writing in "Computer Education" in 1968 argued
very lucidly for involving children with computers. His reasons were threefold, and can be summarized as follows:

1) The growing importance of computers in the wide world.
   The need for an informed attitude towards them.

2) The great future demand for trained computer personnel.
3) Studying computers stimulates creative and logical thinking.

It is clear that he considers the third reason to be the
most compelling of all. Indeed the first two in isolation
would hardly be sufficient. The oil and motor industries
are doubtless equally important to our everyday lives, and
economically even weightier than the computer industry. However there is little inducement to give children more than a
casual acquaintance with them at school. The difference surely lies in that we see the computer as a tool of the mind, a
sort of mental Meccano set, highly instructive to its users,
and capable of varied and open-ended possibilities. The Bellis
report advances a similar argument, and stresses that the
"great majority" of children should learn about computers.
Furthermore the emphasis should not in general be mathematical.
There should be scope for exploiting the computer in various
different disciplines as more people become familiar with
computer concepts. In sharp contrast, the survey of computing
teachers shows that the overwhelming majority are from mathematical departments.

I see the basic objective of computer education as trying
to sow the seeds of logical thinking and of the possibilities
that exist by developing and applying such ideas through a
real computer. This may seem an ambitious task to put across
to the wide spectrum of children, including the unscholarly
ones. I believe, on the contrary, that computing ideas, properly presented, could possibly stimulate an intelligent
interest amongst children otherwise considered unacademic.

Looking at the way the subject has developed in practice it
is apparent that growth has been rather arbitrary and of a
pioneering nature until quite recently. The most notable
successes have occured in schools which, by some means or

other, have acquired their own machine. It is from this
category that come most of the reports quoting the quite
exciting and impressive list of schoolboy achievements, in
the spirit of Mr. Stockwell's third point. (See for example
"Computer Education" No. 1, Sept. 1968, "Second-hand Com -
puter in School", after one year's experience.) The factor
known as "hands-on" (i.e. being able actually to see and
use the machine) emerges as highly important.

The majority of schools cannot afford their own computer,
however, and the available possibilities are principally
either:

> Remote batch-processing with various techniques of
> data preparation and transmission; or:
>
> A remote terminal linked to a time-sharing machine.

Both methods have serious disadvantages for school use. The
first at least is particularly frustrating for beginners
making trivial mistakes and there is no "hands-on" benefit.
A remote terminal in the form, probably, of a teletype has
certain advantages over batch mode. In a limited sense turn-
round is quick, interactive languages are possible, and
there is at least something for the children to see, though
a teletype alone is hardly a machine to capture the imagination. The chief objections are the slow speed of process -
ing, causing a bottle-neck in a situation with numerous
users, and the cost. One school near Cambridge in England
with a link to a commercial time-sharing firm could only
offer 10 minutes per school term to each student, and that
was after 5 p.m. in the afternoon.

There are not many reports to be found on the success of the
distinctly unscholarly children in computing. Here is surely an area where the choice of project and curriculum mate -
rial must be made very carefully. Detailed explanations of

binary numbers and instructions are unlikely to make sense
to a child whose whole environment alienates almost every
academic concept. On the other hand a first demonstration
involves a game of noughts and crosses on a VDU or the analysis of the weights, heights and ages of a class might stimulate a more practical interest. My theme is then that the
introduction to computers should follow lines which would
be valid for any other discipline;

1) The subject should be presented so as to appeal to the
   imagination.
2) It should start with some sound practical exercises
   which will involve the students.

In order to achieve these aims a number of conditions must
be met, basically under two headings: necessary equipment
and suitable syllabus. A third point, presumably in the short
term only, concerns the availability of qualified teachers.

If we consider the objective of giving all or most children
an introduction to computing, it is clear that a substantial
proportion is going to be of quite low academic ability, and
probably unreceptive to anything which they do not find relevant to themselves. The field of computer education pro -
vides an opportunity to break through this kind of barrier,
if only the applications are suitably chosen. The key surely
lies in the peripheral equipment to the computer. A machine
is interesting at first sight when things can be seen and
heard. The more refined intellectual aspect of knowing that
a programme is running or how it is running is unlikely to
appeal to the less gifted children, at least not initially.
The two standard peripherals which match up to this requirement are a fast printer and a VDU. More specialized items
might also be considered. A loudspeaker can give an instructive demonstration of a computer playing music. There have
been other projects which adopt a similar philosophy on peripherals. The LOGO project (4) uses a mechanical turtle on -

line to a PDP 10. The emphasis however has been the more
scientific one of investigating the way in which children
learn to think mathematically, rather than of encouraging
widespread participation. Nearer to the present approach
is Papert & Solomon's article, "Twenty Things to do with a
computer". (5)

Other requirements of a computing service for children are:

> Low and high level language available;
> No data preparation bottleneck;
> Quick turn-round of programmes;
> Good software back-up.

It seems to me that from most points of view it is desirable
to teach in the "hands-on" manner. But of course, not every
school even in a wealthy country, can afford its own com -
puter. I therefore proceed to consider the possibilities of
a mobile computer which could travel round a sizeable area
and provide the necessary service to a whole group of schools.
To give some concreteness to the proposals I have costed such
equipment at current English prices. There would naturally
be some variation from country to country, but I think the
British figures give a fair picture of the magnitude of the
costs involved.

The system I have in mind to meet the requirements consists
of a mobile mini-computer and peripherals as follows:

Item	Cost £	Weight (lbs.)
CPU + 8K store	3,150	50
Fast character printer	2,500	118
Mark sense reader	1,400	75
VDU and keyboard	1,350	35
2 cassette handlers	2,100	52
	10, 500	430

I am not concerned here with advocating the merits of a
particular manufacturer's equipment. There are several which

could provide most of the above items, and the weights and prices refer to products actually available on the market in Great Britain.

The idea of transportable computers is not in itself new. One of the early Elliott 903 computers was installed in a large van used for giving presentations to prospective customers. The machine however stayed in the van. There is a reference in "Computer Education" of June 1971 to a different approach, in which a central processor and a high-speed reader were to be transported between a number of schools, each possessing its own teletype.

The approach to the question of computing teaching is fundamental to this whole discussion. The equipment listed above is extremely flexible, in that it allows the computer to be used as a medium for demonstrating programmes (especially powerful if a cassette operating system is available), or as a local batch-processing system servicing a classroom of children. Mark-sense cards enable each individual to do his own coding, so there is no bottleneck at a teletype or card punch. All the equipment is reasonably light and robust, and could be moved into the lecture-room on a day-to-day basis.

The subject matter of the course should, in my opinion, be chosen very much with the prospective class in mind. A very small amount of programming knowledge would be required to enable an undistinguished class to learn how to construct computer pictures. This idea could be demonstrated on the VDU then switched to the printer. One could imagine the success of a backward class which could produce a picture of the school with a calendar in one term's computing. Playing games by computer is an old favourite and could obviously be especially effective with a VDU. A varied set of peri - pherals in fact affords a great opportunity for exploiting the computer in non-mathematical applications. It breaks

away from the single terminal in the Maths Department
approach. It allows practical exercises to be performed by a
class of students each preparing his own programme, and
receiving diagnostic etc. It can be used in a natural way
to illustrate the link between learning about the computer
and using the computer as a tool. The step from a home -
produced calendar to a reasonable understanding of CBL
techniques is a natural one. This approach provides the
"hands-on" advantage in a very special and complete way.

The uses of such a system would not by any means be limited
to teaching the less gifted children. It could be an ex -
tremely powerful tool in influencing teachers in the use
of computers in the different subject areas. The Bellis report mentions that teachers in general were most interested
in the effect of computing in their own areas. In order to
get them to enlarge their training on computing it seems
necessary to have exercises and packages available which
specifically apply to the various disciplines. This is a
rather artificial situation; deliberately looking for new
uses for a computer simply because the machine is available.
Nevertheless the mobile approach presents attractive possibilities. An external specialist could give a series of
special classes using carefully chosen material, to illustrate the use of a computer in a particular field. I have
in mind a situation in which a considerable amount of programming or data collection may have been invested, and could
be available through a set of cassettes. Initial work could
be done either in educational research centres, or fed back
from the more advanced schools. The mobile computer is serv-
ing here as a convenient medium through which to convey the
ideas. Precisely what the nature of the various applications
might be I am not attempting to forecast. The Bellis report
itself comments rather lamely that computers are not much
used in history, but hopes that in time teachers may be
able to help develop new techniques. I am suggesting that

in order for there to be any chance of this happening then
the computer image must become less formidable to the non-
specialist. The mobile approach seems to be a step in this
direction.

What of the costs of this suggestion? A capital outlay of
about 12,000 Pounds could be written off at about 3,000
Pounds p.a. Maintenance should be of the order 1,000 Pounds
p.a. and running costs another 500 Pounds, giving an annual
expenditure of about 4,500 Pounds. The machine could in
principle be very fully utilized, both in and out of term.
A usage level of 40 hours per week for 45 weeks a year would
give a nominal cost of 2,50 Pounds per hour. True costings
of any proposed scheme are, however, difficult to obtain.
The organization of the teaching support would have to be
considered, as well as the back-up investment for supporting
applications. However, these considerations apply to any
other method of teaching about computers. Other differences
arise because the service can vary in kind, and because the
costs may be borne by different parties. One of the cheapest
methods, from the point of view of the school, is to beg some
free time from local industry ! A scheme at Imperial College
(London) offers a postal batch-service to schools, which
have only the cost of materials and postage to bear. Neither
of these two schemes should be seen as a direct rival to a
mobile computer. On the contrary the role of the mobile computer would be to generate the enthusiasm and understanding
necessary for people to think creatively and naturally about
how to use computers, and hence to obtain sufficient command
of the subject to be able to make effective use of a remote
postal service.

Concerning the question of costs generally it is clear that
if computer education in school is to be taken seriously
then a considerable amount of money will have to be allocated
to this end. My earlier reference to the experience of a
school with a commercial terminal link is perhaps unfair,

because the school was attempting to pay for it from the
existing maths or science budget. The financing must clearly be done at a higher level than the indivudual schools.
With so many computer educational projects under way in the
U.K. at present it is not yet clear whether a general pattern,
as far as equipment is concerned, will become established
and recognized as the best, In a survey of American experience
(6) Mr. James Bailey recommends the possession of a mini -
computer as preferable and cheaper than renting terminals.
The mobile concept should provide a smaller capital invest -
ment on which to work towards this situation, and could play
a useful role in helping to clarify thinking about different
approaches.

As a first step a mobile computer project could be set up
under the control of an educational computer centre so that
the results could be carefully monitored. Several schools,
and possibly also a College of Education should be involved,
and most essentially children with different academic standards. The money spent on such a project would have a double
benefit of providing computer education and research experience into these ideas.

To summarize, I should like to reiterate the key points:

1.   The mobile computer project is proposed as a suitable
     way in which to provide the appropriate computer fa-
     cilities to enable all children to learn about computers.

2.   It could be especially useful in teaching pupils not na-
     turally inclined towards scientific or mathematical think-
     ing, as well as the less bright children generally.

3.   It would be valuable in encouraging teachers to involve
     computing in new fields.

4.   Finally, if I may specialize to the statistical field, it
     would be invaluable in training both those who are in the
     future to become statisticians and those who are going to
     use and interpret statistical material.

References

(1)   Hartley, J.R., Lovell, K., Sleeman, D.H., University of Leeds Research Council's Computer Based Learning Project (Final Report). 1972

(2)   Tinsley, D., A Matchbox Computer. Computer Education No. 6, Nov. 1970, p. 20

(3)   Bellis, Curriculum Papers No. 11, Computers and the Schools. HMSO 1972

(4)   The Logo Project. Bolt Beranek & Newman Inc., Tech. report 1969

(5)   Seymour Papert and Cynthia Solomon, "Twenty Things to do with a Computer". MIT 1971.

(6)   Bailey, James, D. "Survey of the Uses for Schools Computers". Computer Education No. 13, March 1973

THE TEACHING OF PROBABILITY AND STATISTICS AT THE SCHOOL LEVEL - AN INTERNATIONAL SURVEY

L. RÅDE

Chalmers Institute of Technology, Gothenburg

1. Introduction

There is today a very great interest in the teaching of mathematics at the school level. Numerous conferences dealing with the problems of mathematics teaching are arranged, among these the new sequence of international conferences on mathematical education (Lyon 1969, Exeter 1972). New journals especially devoted to this subject has appeared, (1) and (2). Mathematics teaching now also has its own review journal (3).

This interest in the teaching of mathematics has also spread to the teaching of probability and statistics at the school level. This is natural as one of the characteristics of the "new mathematics" movement has been to include probability and statistics in the school curricula in mathematics.

At a conference with many statisticians present it is appropriate to supply this with some statistical data. A biblio - graphy on the teaching of probability and statistics was published recently (39). It contains 132 papers (mainly from the years 1969 - 1972) of interest for teachers of probability and statistics at an elementary level. In the Zentralblatt für Didaktik der Mathematik for 1972 there are reviews of 47 papers dealing with the teaching of probability and statistics. (There were 184 reviews of papers dealing with the teaching of geometry.)

Probability and statistics has also been introduced in the school curricula in many countries. This is for instance the case in Austria (4), Australia (17), (27), Belgium (44), France (25), Great Britain (44), Japan (44), Polen (41) and the Scandinavian countries (38). Also many experiments dealing with the teaching of probability and statistics at the school level are undertaken.

The aim of this paper is to give summary reports on some programs. It is not possible to mention and review all such programs in the field of teaching probability and statistics at the school level. Those discussed have mainly been chosen because information concerning their work has been available. With regard to the many activities in France reference is given to the paper by P.L. Hennequin presented to this con - ference (26).

2. Some school curricula in probability and statistics

Many countries have introduced probability and statistics in their school curricula in mathematics. It is to be expected that these curricula are not so ambitious as are some experimental programs. Table 1 gives a comparison of the contents of some such curricula. The courses listed and the place where information concerning their content has been obtained are as follows.

France: Program suggested by Commission "Lichnerowitz", premiere and terminale class B. P. L. Hennequin, Teaching of probability and statistics in the French lycée, (25)

West Germany: Vorläufige Richtlinien für den Mathematikunterricht in der Studienstufe. Herausgegeben vom Kultusministerium des Landes Schleswig-Holstein. Zentralblatt für Didaktik der Mathematik 1972: 4, pp. 155-162

Great Britain: Syllabus B, Advanced Level. Mathematics Education in Europe and Japan. Secondary School Mathematics Curriculum Improvement Study, Bulletin 6. Teachers College, Columbia University, New York, 1971, (44)

Japan: Curriculum described in Mathematics Education in Europe and Japan, (44)

Sweden: Program for the Gymnasium, Social and Economic Line. Swedish National Board of Education. See also L. Råde, On the teaching of Probability and Statistics at the pre-college level in Sweden, (38)

3. A sample of programs

<u>Working Group on the Teaching of Probability and Statistics at the Second International Conference on Mathematical Education in Exeter.</u>

One of the working groups at the second international conference on mathematical education in Exeter in 1972 was devoted to the teaching of probability and statistics at the school level. A report on the work of the group was written by D. Kaye and F. Mosteller. (32). At one of the meetings plans for the future and possible international cooperation was discussed. The group especially stressed the need for an international news bulletin giving information on experiments and research in the field of teaching probability and statistics.

<u>The Joint American Association - National Council of Teachers of Mathematics Committee on the Curriculum in Probability and Statistics</u>

This committee was formed in 1967 as a combined effort by ASA (American Statistical Association) and NCTM (National Council of Teachers of Mathematics) to deal with the teaching of probability and statistics at the secondary school level. The work of the committee has been presented in reports by W. Kruskal at the CSMP - conference in Carbondale (33) and by F. Mosteller at the International Statistical Institute Round Table in Oisterwijk (35). The work of the committee will also be discussed at this conference (34).

The committee decided to deal mainly with the teaching of statistics at the secondary school level. In this field the committee has started and brought to successful end two projects. One is the producing of a series of four books (each with a teachers manual) with the title <u>Statistics by Examples</u>, (36). Each of these volumes contains examples based upon real-life data, which bring the reader in contact with the methods of thinking of statistics. The books can be used by teachers and students for regular class room study, group

project, self study of teachers in service training. These
books are rich sources for good ideas how to teach statistics at the secondary school level.

The committee has also produced a book "<u>Statistics - A</u>
<u>Guide to the Unknown</u>". (45). This book contains 44 essays,
which discuss on a non-technical level specific applications
of statistics. The anticipated audience is that of people
in charge of secondary school curricula, teachers, and the
general educated public.

The Comprehensive School Mathematics Program

The CSMP (Comprehensive School Mathematics Program) is a
mathematics curriculum project situated in Carbondale, Illinois, USA. Director is Burt Kaufman. General information
about the program is available in (6). It is one of the
program activities of CEMREL, St. Ann, Missouri, USA.

CSMP has since its start 1967 shown a great interest in the
teaching of probability and its applications in statistics
and other fields. In order to get input for its work CSMP
has arranged a series of international conferences devoted
to the teaching of various parts of mathematics at the pre-
college level. The first of these conferences dealt with
the teaching of probability and statistics and was held in
Carbondale in March 1969. The proceedings of the conference
has been published (40). They contain papers by E. F. Becken-
bach, C. B. Bell, H. Davies, A. Engels, H. Freudenthal, J.
Gani, S. Goldberg, B. Kaufmann, W. Kruskal, D. V. Lindley,
J. Neyman, C. R. Rao, A. Rényi, L. Råde, J. B. Douglas,
P. L. Hennequin, B. Hume and S. Holm. The contributions by
A. Engel and C. R. Rao have later also been published else-
where, (19) and (37).

There are mainly two components in the work of the CSMP project. The first of these components is a program for a
broad spectrum of students in grades K-6. This program starts

with a combination of teachers led activities and indepen -
dent work in the earlier grades and later blends into a program comprising multimedia learning materials and involving
a variety of modes of instruction. In this component probability and statistics have been integrated as an essential
part of the program through all school years. The characteristics of the approach and the curriculum detailes have been
explained in (7). Let it here only be mentioned that statistical ideas are introduced already in grade 3, where the
pupils in an activity package organize data obtained by trials
of random experiments and by simulating and also make decisions on data in situations involving uncertainty.

The second component of the CSMP project is a program for
students in grades 7-12. In its present stage of development
it is a program especially designed for highly verbal and
mathematically inclined students. One of the experimental
aspects of this component is to explore the upper content
limits of the mathematics that such students can understand
and appreciate. In this component CSMP has produced a se -
quence of books called the Elements of Mathematics. Of these
(8) - (15) deal with probability and statistics or related
topics. The book (16) is a special version of (8), which
together with five films by A. Engel is an introductory
course in probability for use in grade 7 or 8. The course
takes a little more than 20 hours.

Institute for the Development of Mathematical Education in
the Netherlands

This institute was established in 1971 under the leadership
of Prof. H. Freudenthal. It is involved in various activities
concerning among other things curriculum development in mathe -
matics, school experiments and teacher training. The activities
of the institute are described in the brochure (28), written
in English.

The institute has introduced probability and statistics in its curricula and school experiments both at the primary and secondary level, (30) and (29). The institute has also written the material for a television course in probability called "Kijk op Kans" (vision on chance), (31). It is a program for pupils at the primary level, 10-12 years old. The course material consists of 6 television programs for students, 8 programs for teachers and textbooks (31).

The institute is also publishing a bulletin, Wiskobas Bulletin, (30), which contains among other things descriptions of the institutes work in the field of teaching probability and statistics.

Secondary School Mathematics Curriculum Improvement Study

The SSMCIS (Secondary School Mathematics Curriculum Improvement Study) project is a mathematics curriculum project at Teachers College, Columbia University, New York. Director is H. Fehr. The aim of the program is to create a curriculum in mathematics and the corresponding teaching material, which will break down the barriers between arithmetic, algebra, geometry and the calculus. The project has produced a sequence of textbooks for grades 7-12 of the US high school, (42) and (20). Probability and statistics are taught in school years of the curriculum.

The students in the SSMCIS program can in the last year study a special subject for up to three months. One of the subjects chosen for this study is statistical inference, for which a booklet has been written (43). The emphasis in this course is on point and interval estimation. This is done in connection with the exponential distribution and the Poisson process and not in connection with the normal distribution. Also simulation and experimental design are discussed.

A supplementary list of programs

There are many other experiments and activities in the field of teaching probability and statistics at the school level. The following is a short list of such programs supplementing the list of programs above.

1) The Royal Statistical Society, the Institute of Mathe - matics and its Applications and the Mathematical Asso - ciation (all in Great Britain) has set up a steering committee for the formation of a joint study group whose object would be to improve the teaching of statistics at all non-degree level. Chairman is J. H. Durran from Winchester College.

2) The School Mathematics Project (SMP) in Great Britain has included probability and statistics in their text-books. There is also a special volume giving an alter-native course (18).

3) Work with teaching probability concepts to young child-ren has been reported by T. Varga in Hungary (46) and by E. Fischbein in Roumania (21), (22).

4) The Belgian Center of Pedagogy of Mathematics in Brussels has under the direction of Frederique and Papy done ex - periments in the teaching of probability and statistics. Concerning Frederique's approach in the primary school see (44) and reports in the books (24). An approach for the secondary school is given in (5).

3. Concluding remarks

It is clear from the descriptions above of curricula and experiments that much more has been done with regard to the introduction of probability than with regard to sta - tistics. This may be due to the greater difficulties with regard to the introduction of statistics at the school level but may also be due to the often stated opinion that a course in statistics must be preceeded by a strong course in probability. The last principle might be true at the university level, where this principle, however, is not always followed. But at school level the principle might be questioned. The work of the ASA-NCTM committee, the single really strong experiment in doing something about the teaching of statistics at the school level, indicates that it is possible to introduce statistical ideas without too much preparatory work in probability.

References

(1)   Educational Studies in Mathematics (editor: H. Freudenthal). Published by D. Reidel, Dordrecht, Netherlands

(2)   International Journal of Mathematical Education in Science and Technology (editors: A.C.Baipai and W.T. Martin). Published by John Wiley & Sons, Chichester, England

(3)   Zentralblatt für Didaktik der Mathematik (editor: H. Wäsche). Published by Ernst Klett, Stuttgart, West - Germany

(4)   Lehrpläne für den Unterricht an den österreichischen allgemeinbildenden höheren Schulen (Communication by Dr. L. Peczar)
and Lehrpläne für höhere technische und gewerbliche Lehranstalten (Communication by Dr. H. Aigner)

(5)   H. Breny, Théorie des Probabilités y compris l' Analyse Statistique. Collection Frédérique no. 4. Presses Universitaires de Bruxelles, 1969

(6)   Comprehensive School Mathematics Program, 610 E.College, University City Complex, Carbondale, Illinois, USA 62901 Basic Program Plan

(7)   Guidelines for the CSMP K-6 Curriculum in Probability, Statistics and Combinatorics, 1973

(8) Elements of Mathematics Book 0:8, <u>Introduction to Pro-</u>
     <u>bability</u>

(9) Elements of Mathematics Book 0:12, <u>Topics in Probability</u>
     <u>and Statistics</u>

(10) Elements of Mathematics Book 0:16, <u>Introduction to Com-</u>
     <u>puter Programming</u>

(11) Elements of Mathematics Book 11: <u>Finite Probability</u>
     <u>Spaces</u>

(12) Elements of Mathematics Program Book 12: <u>Introduction</u>
     <u>to Measure Theory</u>

(13) Elements of Mathematics Supplemental Book A: <u>Short</u>
     <u>Course in Mathematization</u>

(14) Elements of Mathematics Supplemental Book B: <u>EM Problem</u>
     <u>Book</u>

(15) Elements of Mathematics Supplemental Book C: <u>Computer</u>
     <u>Oriented Mathematics</u>

(16) <u>A short course in Probability</u>

(17) J.B. Douglas, <u>On the teaching of probability and statis-</u>
     <u>tics at the pre-college level in Australia.</u> The Teaching
     of Probability and Statistics. John Wiley & Sons, New
     York 1970

(18) J.H. Durran, <u>Statistics & Probability</u>, Cambridge Univer-
     sity Press 1970

(19) A. Engel, <u>Teaching Probability in Intermediate Grades,</u> Int.
     J.Math.Educ.Sci.Technol. vol.2:3, 1971, pp. 243-294

(20) H. Fehr - J.T. Fey - T.J. Hill, <u>Unified Mathematics,</u>
     Course 1-3. Addison-Wesley Publishing Company, Menlo Park,
     USA, 1972

(21) E. Fischbein - I.Pampu - Minsat, <u>Initiation aux probabi-</u>
     lités à l'école élémentaire, Educ,Stad. in Math.2, 1969,
     pp. 16-31

(22) E. Fischbein et alii, La Nature Psychologique du Concept de Probabilité et l'éducation Intellectuelle, Revue Romaine des Sciences Sociales, Serie de Psychologie 13, 1969, pp. 155-166

(23) Frédérique, Première Leçon de Probabilite, Nico 9, 1971 pp. 64-72

(24) Frédérique, Les enfants et la Mathématique, vol 1 -, Didier, Brussels

(25) P.L. Hennequin, Teaching of Probability and Statistics in the French Lycée, The Teaching of Probability and Statistics, John Wiley and Sons, New York 1970

(26) P.L. Hennequin, Tendances de l'enseignement de la statistique en France dans le second cycle des lycées, International Statistical 3rd Round Table on Teaching Statistics at the Secondary Level, Vienna, 1973

(27) B. Hume, The introduction of probability and statistics on the pre-college level in/Western Australia, The Teaching of Probability and Statistics, John Wiley and Sons, New York 1970

(28) I.O.W.O, 1972. Institute for the Development of Mathematical Education; I.O.W.O. Tiberdreef 4, Utrecht, Netherlands.

(29) Waarschijnlijkheidsrekening en Statistiek voor het V.W. O., 1972

(30) Wiskobas Bulletin

(31) Kijk op Kans, Werkboek and Onderwijzerboek 1972.

(32) D. Kaye - F. Mosteller, Report on the working group 5, at the I.C.M.E. international conference on Mathematical Education in Exeter 1972. See also A.G. Howsen, Developments in Mathematical Education. Proceedings of the Second International Congress on Mathematical Education. Cambridge University Press, 1973.

(33) W. Kruskal, Statistical Examples for Use in High
     School, The Teaching of Probability and Statistics,
     John Wiley and Sons, New York 1970

(34) W. Kruskal, Towards Future Activities of the JCCSP,
     International Statistical Institute 3rd Round Table
     on Teaching Statistics at the Secondary Level, Vienna
     1973

(35) F. Mosteller, The Joint American Statistical Associa-
     tion - National Council of Teachers of Mathematics
     committee on the Curriculum in Statistics and Proba-
     bility, Review of the International Statistical Insti-
     tute 39:3, 1971, pp.340-342

(36) F. Mosteller - W. Kruskal - R. Link - R. Pieters - G.
     Rising (editors), Statistics by Examples, Vol.1: Ex-
     ploring data, Vol.2: Weighing Chances, Vol.3: Detec-
     ting Patterns, Vol.4: Finding Models, Addison-Wesley
     Publishing Company, Menlo Park, USA, 1973

(37) C.R. Rao, A Multidisciplinary Approach for Teaching
     Statistics and Probability, Int. J. Math. Educ. Sci.
     Techno. vol.2:3, pp. 295-312, 1971

(38) L. Råde, On the teaching of probability and statistics
     at the pre-college level in Sweden, The Teaching of
     Probability and Statistics, John Wiley and Sons, New
     York 1970

(39) L. Råde, A Bibliography on the Teaching of Probability
     and Statistics, Zentralblatt für Didaktik der Mathema-
     tik, 2/72, pp. 70-72, 1972

(40) L. Råde (editor), The Teaching of Probability and Sta-
     tistics, John Wiley and Sons, New York 1970

(41) W. Szlenk, Some remarks on teaching of probability theory
     in Poland, personal communication

(42) Secondary School Mathematics Curriculum Improvement
     Study, Teachers College, Columbia University, New York
     Unified Modern Mathematics, course 1-6

(43) L. Råde, P. Smith, <u>Introduction to Statistical Inference</u>.

(44) <u>Mathematics Education in Europe and Japan</u>, SSMCIS Bulletin 6

(45) J.M. Tanur (editor), <u>Statistics: A Guide to the Unknown</u>, Holden Day, San Francisco, USA, 1972

(46) T. Varga, <u>Logic and Probability in the Lower Grades</u>, Educational Studies in Mathematics, 4, 1972, pp.346-357.

Table 1. Comparison of content in some high school curricula in probability and statistics

	West Germany	Japan	France	Sweden	GB
Finite outcome sets	x	x	x	x	x
Interval as outcome sets	x	x	x	x	x
Location characteristics	x	x	x	x	x
Dispersion characteristics	x	x	x	x	x
Normal distribution	x	x	x	x	x
Hypothesis Testing	x	x	x	x	x
Expectation	x	x	x		x
Denumerable outcome sets	x		x	x	x
Variance	x	x	x		x
Estimation	x	x	x	x	
Binomial distribution	x	x		x	x
Confidence Interval	x	x	x	x	
Combinatorics	x	x		x	
Random Variables	x	x	x		
Poisson distribution	x			x	
Axioms of probability	x		x		
Conditional probability	x	x			
Independence		x		x	
Chebyshev inequality	x		x		
Regression	x		x		
Weak law of large numbers	x				
Random numbers and simulation		x			
Markov Chains	x				

D I S C U S S I O N S

J. Oyelese: In West Africa, both the Joint School Project, which is sponsored by the Ghana Mathematical Association and the Entebbe Mathematics Project, which is sponsored by the Educational Development Corporation of Boston, Massachusetts, USA, have introduced statistics in their textbooks. While the Entebbe Project has been produced for experimental use in English speaking African countries, in East and West Africa the Joint School Project have been produced in the first place for schools in Ghana even though they are now used in other countries in West Africa. The Entebbe Project has also devoted some space to elementary ideas of probability and statistics.

FIRST STEPS IN TEACHING PROBABILITY

M. HALMOS

Mathematical Institute of the Hungarian Academy of Sciences,
Budapest

1. Introduction

I could not start more adequately than by quoting Alfred
Rènyi about what he called the three main aims of the teaching of probability (1):

" 1. Probability should be taught because it is important for
the mental development of the students.
2. Probability should be taught for its practical uses in
everyday life and in different fields of knowledge.
3. Probability should be taught because its teaching is an
important and even indispensable part of mathematical
education."

My main concern in what follows will be point 1 of the above:
probability for mental development. At the first steps of
teaching probability this is perhaps the most decisive aspect.
The study of probability supposes a special way of thinking
that is unknown and strange for the students. Therefore the
first step is to show this unknown "world": What does " it is
a matter of chance" mean ? What is the nature of the statis -
tical regularities in everyday life, in nature, in games of
chance, etc.? I do not mean that the teaching of probability
should start with the theory of statistics. What I do mean is
that a course in probability should start with well-chosen
games, experiments, and examples. These should make students
familiar with chance, with statistical ideas and should make
them ready for some fundamental probabilistic notions. Then
only can we start with the theory of probability and, later,
of statistics.

In the following I shall describe some experiments carried
out in 4 classes with twenty-odd or fewer 14-15 year old

students in each. Their main field of interest has been mathematics.

2. The random world

a) The teacher presents two sequences of zeros and ones, and
tells the students that one of them was obtained by throwing a coin (O means heads and 1 tails) and the other one is
a fake. It is only an imitation of a real random sequence,
produced by a pupil. The students have to guess which is a
real and which is not.

Here are two sequences Which is the random one?

```
01011001100101011011010001110001101101010110010001 1
01010011000110101100101100101100100101110110011011 1
11010010110010101100010011010110011101110101100001 00
11000100010011101001011000110111001010001101100110 0
```

```
100111011111010011100100111001000111011111101010101
111000000100010100100000010001100010100000000011001
0000000111110000110101001001001111110100110001100 0
10001111100011110101000101001110111100111011110011
```

You will have guessed by now that the second is the random
one. Most students believe, however, that the first one is
the real random sequence. They imagine that a proper random
sequence cannot contain long runs of either zeros or ones.

b) A student throws a die several times. The question is:
How often will a number less than 3 appear? They guess
that this would happen in about one third of the throws.
They also foretell that in many throws there will be
some sort of "convergence". Though this is not far from
being true, it is important to make clear that this is
not the same as the usual convergence. In any number of

throws, as large as you want, deviations may exceed any number. It is worth while to represent the data on a graph paper to show the tendency of relative frequency - here and also in the other experiments.

There are still many misunderstandings about randomness. One should carry out further experiments to get free of these errors. But I shall not continue this strand now.

3. Independence

In our classes each student has a table of random numbers. They use it to simulate experiments, among them the follow - ing:

a) One of the two students look up ten random digits in his table, e.g. ten consecutive digits starting from a randomly chosen row and column. The other has to guess all of them one after the other. He is not allowed to ask questions such as "Is it even?" "Is it smaller than five"?, or anything like that. He can only suggest that the first digit among the ten is, say, 8, and the only information he can obtain is "yes" or "no". So it goes on until he guesses the first digit correctly. Then they pass to the second digit. And so on up to the tenth.

This is not particularly challenging as a game. More interesting is the discussion which follows it:" What is the best strategy?"

When I tried out this game in a class the students entered into a great controversy about this question: Some of them said: "All strategies are alike. You suggest different digits until you hit upon the correct one". The digits not yet found should be suggested first". If, for instance, the digit 6 has been found to be the first in the row, then 6 should be the last guess for the next digit.

Investigations of the frequency of digit pairs 00,01,02,etc. helped them to reject the wrong idea. They also investi - gated such digit sequences starting with 222. Such investigations led them to the feeling that none of the pairs or triplets or, for that matter, any multiplets or digits is distinguished. They arrived at the conclusion that after any number of 2's a further 2 was not less unlikely than any other digit.

Their formulation of this property became:" In the table of random digits no digit could remember the digit preceeding it." This was also found to be the property of random digits produced by themselves, e.g. by throwing dice or coins.

Other variations of the above games are:

b) Instead of ten random digits 10 random letters were to be guessed. Randomness was generated by selecting every 30th letter in a text.

c) The same with 10 consecutive letters of a text.

The distribution of the letters of the alphabet was found to be rather uneven both in b) and c). My students found a good strategy for b) to be: suggest first the most frequent letter, then the next frequent, etc.. They would accept this strategy in game c) as well, but then they realised that letters in this case were not independent:" they remembered" the letters preceding them.

4. Expectation

a) Ten cards are given with the digits 0, 1, 2, 3, 4, 5, 6, 7, 8, 9 written on them.

One of two players puts the cards in a hat, mixes them, and draws one of them without showing it to the other players. The second player, as in the preceding games, tries to guess

the digit on the card. They record the number of trials ending in a correct guess. (This might have been done in the
preceding games as well.) Then the first player replaces the
card and draws a new card, after mixing them thoroughly. They
do this a certain number of times. At the end they count the
average numbers of trials. The problem is: in the case of a
given strategy what is the expected count? Students find this
similar to 3 (b).

b) Same as above, but with following cards:

In a) there is no best strategy (or rather, all strategies
are best). The students, e.g. decide the strategy to be:
the second player suggests the digits in the natural order:
0, 1, 2, ...9,. Suppose, first player draws n cards, and
second player guesses each of them successively, before the
next is drawn. What is the expected number of trials? The
reasoning of the students was something like this: If the
first player draws the card 0, the number of trials is 1.
If he draws the card 1, the number of trials is 2, and so on.

The number of trials may be: 1, 2, 3, ...10. In about 1/10
of all the cases the number of trials will be 1, as in about
1/10 of all cases 0 is drawn. Similarly in about 1/10 of the
games the number of trials will be 2, and the same is true
for 3, 4, 5, 6, 7, 8, 9, 10 trials.

This reasoning gives for the expected number of trials:

$$\frac{\frac{n}{10}\cdot 1 + \frac{n}{10}\cdot 2 + \ldots + \frac{n}{10}\cdot 10}{n} = 0.1\cdot 1 + 0.1\cdot 2 + \ldots + 0.1\cdot 10 = 5.5$$

In the case of game (b) the best strategy is: the second

player suggests the digits in the following order: 1, 0, 5.
We will denote it by "strategy 105". Similarly we can speak
of "strategy 150", "510", etc.

In the case of strategy 105 the expected number of trials is

$$0,5 . 1 + 0,3 . 2 + 0,2 . 3 = 1,7.$$

Their reasoning was: in n games 1 trial can be expected about
in n/2 cases, 2 trials in about 3n/10 cases, 3 trials in about
n/5 cases.

This sort of games led them to the idea of expectation. The
students calculated the expectation for other strategies as
well (510, 150 etc.) and found that in the case of strategy
105 the expected count is the least.

An additional question was: what about the student who knows
that digits 0,1,5, and no other digit figure on the card in
the hat, without knowing their proportions? He guesses randomly. What is for him the expected number of trials?

The students found that in this case 1, 2, or 3 trials occur
equally likely, independently of the proportions of the di -
gits 0,1,5, and therefore the expected number of trials is
2.

5. "I am converted"

Somebody looks in a table with 4 drawers for an envelope.
He is not quite sure he has put it into one of the drawers;
he gives a 2/3 chance to have done so. He opens three of the
four drawers. No envelope is found. He is to open the fourth.
What is now the subjective probability of finding the envelope?

The students were ready to answer: 1/6. This number is, of
course, wrong. They did not consider the condition (the fact
that none of the three preceding drawers did contain the envelope). I found it impossible to discuss their wrong answers
with them, because they had not yet met with the idea of con-

ditional probability. This problem was suggested precisely
as a first step towards this idea.

Therefore I did not reject their answers. Instead I repeated
the problem, stressing the condition, once again, remarking
that I wondered whether their answers were correct. They asked
me to check my answer; they checked theirs, they said, and
found it to be correct.

Then we agreed to devise an experiment simulating our problem. They agreed to use a table of random digits, looking up
the digits 1, 2, 3, 4, 5, 6, in it; 1 meaning to find the envelope in drawer first, ... 4 to find it in drawer 4-th;
5 and 6 not to find it in any of the drawers.

They understood that only the digits 4, 5, 6, should be
looked up, because no envelope is in the three of the four
drawers.

While devising the experiment one of the students said loudly:
"Oh, I am converted." And he explained: after having found
the first three drawers empty, the probability of finding it
in drawer four, was bound to be 1/3. It could not be 1/6 any
more.

This was the point when I felt it was time to tackle the
theory.

6. Influences in other fields of mathematics

a) analysis

The students throw a coin and observe the outcome as head or
tails (H or T), then he continues and obtains a series of
heads and tails, for instance HTTTH. He stops when the ob -
served sequence of outcomes is symmetrical for the first
time. He does not stop, however, after two tossings. This
means that he stops after series like HTTTH, THHHT or HTH,
but he does not stop after series like HTTT, THH or HT. What
is then the expected number of tossings, the first toss not
being counted ?

The students notice that the possible number of tossings are
2, 3, 4,... with probabilities 1/4, 1/8, 1/16....To find the
expectations one has to sum the series

$$2 \cdot \frac{1}{2} + 3 \cdot \frac{1}{4} + \ldots + n \cdot \frac{1}{2n-1} + \ldots$$

At this stage the students do not know how to find the sum of
series, but they can perform actual series of tossings and
find the mean of the observed number of tossings. Already
after a small number of such series they will estimate the
sum to be about 3. I think this is a good preparation for
analysis.

b) <u>combinatorics</u>

Let me tell a story about something which happened in a lesson,
where the subject was combinatorics, which took place after
some lessons in probability. The following problem was given
to the students: Prove that

$$0 \cdot \binom{n}{0} + 1 \cdot \binom{n}{1} + \ldots\ldots + n \cdot \binom{n}{n} = n \cdot 2^{n-1}$$

Several proofs were given by the students. One remarked that
he had a very interesting proof, where he used a random walk
situation which we had discussed in a probability lesson.

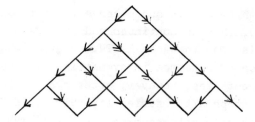

Fig. 1

Consider someone who walks at random in the network in Fig.1. At each corner he goes right-down or left-down with probabilities 1/2 and 1/2 respectively. Number the possible corners after n steps by 0, 1, 2, ..., n according to figure 2.

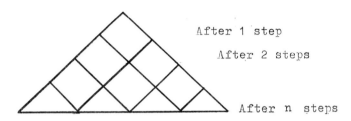

After 1 step

After 2 steps

After n steps

Fig. 2

It is then trivial that the expected number of the corner after n steps is n/2. But this expected number is also

$$0 \cdot \frac{\binom{n}{0}}{2^n} + 1 \cdot \frac{\binom{n}{1}}{2^n} + \cdots + n \cdot \frac{\binom{n}{n}}{2^n}$$

which gives the desired formula.

References

(1) Rényi, A.: Remarks on the Teaching of Probability. In L. Rade (editor), The Teaching of Probability and Statistics. Almquist & Wiksell, Wiley Interscience, New York 1970.

TRENDS IN THE TEACHING OF STATISTICS IN FRANCE IN SECONDARY SCHOOLS

P. L. HENNEQUIN

Université de Clermont-Ferrand, Clermont-Ferrand

1. Programs of Mathematics

French secondary education is organized by national programs elaborated by committees including representatives of teachers, of the universities, and of Administration; when they are published these programs are accompanied by comments written by the Inspector General. These programs are the more con - straining when they concern a form (grade level in school) in which an examination is taken, as is the case of terminal forms which lead to baccalaureate (the last year of secondary school).

We have had occasion to present in (1) the programs of sta - tistics and probability theory used in 1969. Since then, the "Lichnerowicz Committee" has set up new programs theoretically scheduled for 1970 for the first form and the terminal form; in fact, these programs are required only for the baccalaure- ate, and are thus taught only in that school year. In (2) will be found the complete text of these programs and of the comments accompanying them. Let us indicate here some charac- teristics which distinguish them clearly from the preceding ones:

a) the teaching of statistics and probability is now appear- ing in all areas (humanities, economics, experimental scien- ces, technology), in "premiere" and in "terminale" (grades 11 and 12).

b) in the minds of the authors of the program, the teaching of statistics, and particularly of statistical inference, must be preceded by a good introduction to probability theory, at least for the finite case.

The new programs are thus devoted to that study, carried on
up to the weak law of large numbers and the study of binomial
distribution.

The study of descriptive statistics, and especially of correlation, has been much reduced. The notions of linear re -
gression, time series, and economic indices have disappeared
from the program, as well as the study of the Laplace-Gauss
distribution, sampling theory, and estimation of a mean.

One can thus say that now statistics intervenes in secondary
education only in its descriptive part, considered as introductory to probability theory.

The schedule designed for that teaching includes, in "premiere"
as well as in "terminale" (according to the area), between
ten and thirty hours annually, including exercises, which
leaves too little time to learn manipulative skills to study
concrete situations.

Most teachers complain about the extent of the mathematical
programs and in certain Academies have simply proposed the
suppression of that part of their teaching considered by
them to be supplementary and independent of the other chapters.
We should note, however, that some specialized sections of
technical education oriented to management develop statistics
more widely.

2. The role of statistics in the teaching of other disciplines

Although French teaching is rather strictly partitioned into
disciplines, the teaching of statistics is developed outside
of mathematics. Appearing as a supplement in physics in the
chapter about "measurement and uncertainty" or in philosophy
apropos of "statistical determinism" or in the construction
and use of statistics of data in sociology, it is the object
of a real life study in biology in some sections of the
"premiere" form (population and ecology) and in all sections
of the "terminale" form. This study includes the analysis and
exploitation of quantitative results concerning the variation

in a progeny and in a population where a normal distribution is fitted to the empirical distribution. Statistical rules for the transmission of hereditary characters supply good examples of finite probabily spaces. One can then analyse the phenomena of "linkage" and "crossing-over", which do not conform to the independent selection of characters. The conformity to Mendel's laws is tested for fit, using in particular the x^2 - test.

This use of statistics in the teaching of biology can be very instructive and is usually scientifically valuable. It often makes reference to the teaching of mathematics, sometimes before the notions employed appear there explicitly.

One could expect to find statistical developments in the teaching of economics (section B and G). Though numerous statistical tables are quoted there, too few criticisms are available concerning their validity and even fewer precise indications concerning the way they were obtained or constructed.

3. Experiments in the primary school (with children 8-10 years old.)

Although our report is devoted to secondary education, we think it is necessary to say something here about primary education because experiments during the last few years indicate the merits of exposing the child at a very early age to random situations.

Following in particular the work of the CEMREL, divers experiments have begun in France, very often on the initiative of one of the "I.R.E.M." - institutions.

The earliest ones (1970) were started at the school of "Francheville-le Haut" under the direction of the I.R.E.M. of Lyon. The most systematic were organized by G.Brousseau in Bordeaux (of (3)); others have begun in Clermont-Ferrand.

An entire series of experiments has been designed to study the behaviour of a child confronted by situations involving chance: comparison of various modes of representing a series of random drawings, progress towards a decision (relative, for example, to the contents of a bag or a box). Linguists are studying systematically the vocabulary used by a child in this process .

Let us note an other experiment carried out within the framework of the education of a student teachers at the "normal school" and followed by manipulations in the elementary forms; it includes practical and theoretical study of a machine that can be programmed to learn to play well the game "run at seven" (a variant of the game Nim).

This study leads rapidly to interesting and difficult mathematical problems (asymptotic behaviour). It provides at the same time a model of a learning situation. Let us remember the idea of balancing the mathematical and pedagogical training of future teachers.

4. The use of calculating machines and computers

The teaching of statistics can employ:

a) calculating machines which avoid the rapidly tiring aspects of hand calculating (calculation of means, variances, re - gression). However, we must warn the pupil of the ease with which these machines give a large number of decimals, most of which are insignificant, to say nothing of the danger of outright errors.

b) Mini-computers (worth 4 000 to 10 000 U.S. dollars) which have in addition a routine for general random numbers, allowing the rapid simulation of a large number of drawings (unfortunately, most of the routines employed generate arithmetic sequences which are not equidistributed).

c) Computers (worth more than 120 000 U.S. dollars), access-ible, for instance,by a console. One can then envisage simulating the usual processes (Poisson process, Brownian motion, birth and death process, Galton process, Markov chains); writing the generating routines of these processes from a sequence (x_n) of independent uniform variables on (0, 1) is a good exercise.

Experiments in this direction are taking place at the mo -
ment in Grenoble and Clermont-Ferrand. The policy of the
Ministry seems to be, (in order to develop the role of computer science in secondary education) to install such a computer in the principal establishments (300) within a few
years; These computers would be the responsibility of a team
of teachers who have received a year of full-time instruction.

5. Problems posed by the training of teachers

Statistics is still too often pushed aside a bit and treated
as a poor relative with respect to other branches of mathe -
matics; there are several reasons for this:

a) The training of teachers who left the university ten years
   ago did not include this subject.

b) The teaching of statistics should include, beyond mathe -
   matical definitions and theorems, a systematic study of
   concrete examples, the mathematisation of which must be
   done step by step, the posing of different models, and
   their study and criticism after new experiments. These examples should be taken from physics and the experimental
   sciences as well as from geography, economics, sociology,
   and linguistics.

This construction of models, in favor in the Anglo-Saxon countries, is often sacrificed in France because it requires both
time and research exterior to the domain of mathematics.

c) finally,the object of probability theory, the study of
   chance, which is a metaphysical idea more or less familiar,
   requires if it is to be treated as a mathematical program
   (law of large numbers and central limit theorem), the knowledge of advanced tools of topology and analysis unknown
   to pupils of secondary education. This makes its elementary
   presentation difficult, especially if it has not been preceded by any qualitative study of the repetition of trials.

For this reason it has been necessary to undertake the systematic instruction of teachers in this domain. But since, at the same time, it was convenient to initiate the oldest of them to other totally new chapters of mathematics that they had never taught and hardly practised, these efforts, begun about five years ago, have just now borne fruit. In general the initiative in this has been taken by the I.R.E.M., be - ginning in 1968 these institutes have been created at the rate of three per year, and will cover the whole of France by 1974; this initiative has resulted in the production of documents in particular in Montpellier (4) and in Rennes (5).

The comparison of the work done in the I.R.E.M. in probability as well as in statistics, has been made possible by means of two colloquia, the first in Lyon from the 29th to the 31st of May 1972 (6), the second in Clermont on the 30th and 31st of March 1973 (7).

Each national congress of the Association of mathematics professors, in Toulouse in 1971, Caen in 1972 and Nancy in 1973, has included a committee report on this subject.

In December 1972, the Ministry of National Education organized on the initiative of the "Association of University Statisticians" which includes a hundred teachers in higher education, a three-day meeting which attracted 60 participants from all over France; once again this was aimed at secondary teachers, but this time from all disciplines. In fact it concerned especially teachers of economics, and most of the debates concerned the difficulty of gathering statistical information. This experiment is to be undertaken again during the next school year and will include a larger number of specialists.

Addressing a larger public, the "Chantiers Mathematiques" of the school television (OFRATEME) have devoted five of their productions in 1972 and 1973 to probability theory and statistics. Four of these productions of thirty minutes took as their point of departure a concrete problem (learning, selling price of a perishable good, medical diagnosis, telephone exchange); using simplifying assumptions, they proposed a mathe-

matical model employing finite probability spaces, treated
that model, and opened up for each example the possibility
of making decisions at a given significance level. The fifth
production, devoted to the central limit theorem, shows the
convergence of the binomial law to the normal distribution.

Each of these productions is accompanied by a written document (8) of about fifteen pages which allows the teacher both
to see how to show concrete examples to his pupils and to
improve his knowledge with some theoretical developments at
the univerity level.

Conceived by a team of Clermont teachers, these productions
have placed in evidence the advantage of producing, if possible at a low price, short films (5 minutes) which allow teachers
to present to their pupils either sequences of random variables
(or processes constructed from successive and very rapid drawings), or convergence toward the characteristic limit distributions, the time parameter giving an additional dimension of
the graphical representation.

The use of graphic display connected to a computer makes it
possible to write the generating routines of such films very
rapidly, then to realise them by direct filming of the console
screen; this considerably decreases their cost compared with
classical techniques of animation, which require a substantial
team of artists who are not always mathematicians !

6. The prospects

Three tendencies seem to us to favor a rapid evolution of the
situation and a development of the teaching of statistics :

a) Primary instruction is very open to mathematical revival
   and to experimentation with new methods. The close connection between probability and games, together with the
   possibility of group experiments with the participation of
   all, everyone repeating the experiment at his turn, are
   excellent motivations for an experimental approach to chance.
   This approach seems available as soon as the child, at

about 8 or 10 years, begins to be able, on the one hand, to manipulate ratios and especially fractions, and on the other hand, to use various graphic means to represent the result of a series of drawings.

b) A reform of mathematical instructions, begun in the 6th form (first form of secondary instruction) in 1969, will terminate this year in the 3rd. The pupils of the first cycle of the lycees are now familiar with elementary alge - braic structures. One can then envisage (of (9)) formalizing the construction, analysis, and publication of an investigation or survey.

Having the pupils construct a questionnaire, carry out the corresponding inquiry (using as subject the pupils of their school or the inhabitants of their community), and analyze the data, will show both the difficulties of surveying a population and the possibility, once the answers to certain questions are known, of deducing what would have been the response to other questions, had they been asked.

The corresponding mathematical structure for the survey is that of a finite probability space and there are no diffi - culties in expositing, without reference to chance, the no- tions of random variable, of conditional probability and of independence (related to the notion of representative sample).

c) In the second cycle (second, first, and terminal forms), we seem to be orienting ourselves toward a common core supplemented by different specialized options. This would then be the time to complete the construction of probability theory and to introduce the notion of product space (indepen- dent repetitions of an experiment), then to prove the weak law of large numbers and the convergence of the binomial dis- tribution.

One could then show without difficulty how these mathematical tools permit the construction of a model for random situations for which the pupil should have had a good qualitative approach in elementary school and which would thus be familiar to him.

Thus the roadblocks that we meet too often when we introduce
the mathematisation of chance without preparation would be
eliminated or at least reduced.

This teaching could culminate in the explanation of simple
decision strategies, for which the importance of the choice
of significance level would be discussed.

Of course, such an evolution supposes, before its generaliza-
tion, serious experimentation by volunteer teams of pedagogues,
freed from some administrative constraints and from the strict
partitioning into disciplines. This experimentation is, in
France, still much too modest in size in secondary education.

The "Association des professeur de Mathématiques" asked, in
its "Charter of Caen" in May 1972, that these experiments take
place in a sufficient number of schools so that the results
could be compared.

By carrying out this excellent plan, we can at least put statistics, an interdisciplinary domain par excellence, in its
natural central place.

References

(1) The Teaching of Probability and Statistics (edited by
     Lennart Råde), Almquist - Wiksell, Wiley Interscience,
     New York 1970.

(2) Programmes et commentaires B.O.E.N. = Bulletin officiel
     de l'Education Nationale
     Classes de 1ère programmes B.O.E.N. nr. 17 du 23.04.70
                          (rectificatifs nr. 2 et 48, 1971)
                          Commentaires B.O.E.N. nr. 4 du 28.01.71
                          (rectificatifs nr. 7 du 18.02.1971)
     Classes Terminales programmes B.O.E.N. nr. 25 du 24.6.71
                          commentaires B.O.E.N. nr. 30 du 29.7.71

(3) Vers un enseignement des probabilités à l'école élémen -
     taire par une équipe dirigée par N. et G. Brousseau, I.R.
     E.M. de Bordeaux - 1972

(4) Probabilités et statistique (classe de 1ère), I.R.E.M.
     de Montpellier, 1972

(5) Probabilités - I.R.E.M. de Rennes, 1972

(6) L'enseignement des probabilités, colloque inter I.R.E.M.
     Lyon - 29 - 31 Mai 1972, I.R.E.M. de Lyon, 1972

(7) L'enseignement de la statistique, colloque inter I.R.E.
     M. Clermont-Ferrand, 30-31 Mars 1973, I.R.E.M. de Cler-
     mont-Ferrand, 1973

(8) Chantiers Mathématiques 1972-73, Tome 2, O.F.R.A.T.E.M.E.,
     1973

(9) Hennequin, P.L.: Pour un enseignement de la statistique
     dans le second cycle - Bulletin de l'A.P.M.E.P. (à
     paraître)

Acknowledgement

I would like to thank Robert Smythe for his kind help in trans-
lating my paper into English.

TOWARDS FUTURE ACTIVITIES OF THE JCCSP

W. KRUSKAL

Department of Statistics, University of Chicago, Chicago

We have had a discussion of the formation of the Joint Committee on the Curriculum in Statistics and Probability, and we know about its two publications,

Statistics: A Guide to the Unknown, and

Statistics By Example.

In the present discussion, I shall outline possible future activities of the Committee. The spirit will be one of seeking suggestions and criticism; in any case, I can give only a personal view in these notes since they have been perforce prepared before the next meeting of the Committee in May.

There are recent changes in the membership of the Committee, and I list here its present roster,

Richard Pieters (Chairman), Phillips Academy, Andover, Massachusetts

Max Bell, Department of Education, University of Chicago

Richard Light, Graduate School of Education, Harvard University

Gottfried Noether, Department of Statistics, University of Connecticut

Albert Shulte, Supervisor of Mathematics, Pontiac, Michigan

and myself.

These introductory remarks have centered about the Joint Committee, but I do not want to give a parochial misimpression. The Committee has no monopoly of experience, energy, or idea. I hope that the activities of the Committee, past and future, will suggest similar or better activities for other indivi duals and groups. In short, I hope that the discussion today

will center on content rather than on organizational affilia-
tions or procedural details.

More of the same --- or of the similar.

Initial indications are that both our publications are well
received and are becoming widely known and widely used. Indeed,
the uses were at first a little surprising to me. For example,
Statistics: A Guide to the Unknown, intended originally as an
exciting introduction to statistics for the general public,
is probably being used mostly as motivating auxiliary reading
in statistics courses.

Hence a natural possible future activity would be the crea -
tion of A Further Guide to the Unknown or Statistics By Additional Example. Such initiatives would be quite feasible; in
particular, a number of ideas for essays in the Guide, and for
examples, were not used for reasons having nothing to do with
their intrinsic merit. So we would be starting with a back -
log of specifics.

Yet I doubt that the Committee itself will try to climb the
same mountains twice. First of all, we no longer have the
stalwart leadership and climbing ability of Frederick Mosteller; the leadership and abilities of Richard Pieters may well
move us toward other peaks. Second, we have now shown that
the mountains can be scaled, that it is indeed possible to
write nontechnically, accurately, and interestingly about statistics, and to furnish examples at once realistic, relevant,
and capable of treatment at many levels of background.

There are, however, some other important directions or initiatives that come under this general category. In particular ,
examples based on official statistics - so often wrongly regarded as dull and meaningless - might form the basis of a
special activity. The sort of example I have in mind is al -
ready represented in Statistics By Example: my own example
called "Babies and Averages" is one; another is Bradley Efron's

example on "Black and White Survival in the United States", and a third is Joseph Sedransk's "Prediction of Election Returns from Early Results". These could be multiplied manyfold to bring in other kinds of official statistics, to show inter-relationships among them, and to indicate the existence of fascinating intellectual questions together with clearcut immediate practical relations with the day to day world.

A second kind of special direction of extension might be publication of example sets dealing with designs and data from elementary laboratory courses. There is at least one such in Statistics By Example, "The Acceleration of Gravity", by John Mandel. It deals with measurements on the lenght and periods of simple pendulums in an elementary physics laboratory course at a Netherlands university.

It may well be that school and college laboratory work represents a substantially untouched field for useful statistical discussion. The late W.J. Youden was concerned with these matters, and one may see something of his views in the book he wrote with secondary-school use in mind: Experimentation and Measurement, Scholastic Book Service, New York, 1962, produced by the National Science Teachers Association and the National Bureau of Standards.

So there are two possible extensions of Statistics By Example, both framed to bring out greater variety and intensity along specific lines than the precursors that have already appeared.

Variations on a theme.

Here I list first initiatives toward translation into other languages than English. We have taken some small steps in that direction, and comments from this highly international, linguistically broad, group would be most welcome.

It has been proposed that Statistics By Example be reorganized and amplified to put it into more traditional textbook form. That strikes me as a fine proposal that turns on finding one or two dedicated, able authors for the work.

There have been interesting suggestions for special course
materials based on Statistics By Example. These range from
materials for a twelfth grade course for good college-bound
students to a ninth grade course for non-college-bound
youngsters.

It has also been suggested that there be more explicit treatments that tie the statistical examples to the standard mathematics curriculum --- and perhaps vice-versa.

Another direction - one that is already begun in Statistics
By Example - is linkage of the examples with numerical work
that can be done on small computers or calculators of the
kinds that are becoming more and more widely available. The
linkage might require only an inexpensive calculator that
handles basic arithmetic, or it might call for a small pro -
grammable computer. To see what the latter might look like,
we may examine Martha Zelinka's Example, "Predicting the Outcome of the World Series", in Statistics By Example. That particular case refers to American baseball and thus might not
be very effective in other countries, but it suggests a kind
of highly useful teaching device that could be used in more
nearly universally relevant contexts. As another example, I
cite the fascinating approaches of A. Engel, which I heard
about in his talk at the Philadelphia AAAS meeting.

Another sort of continuation would be encouragement of the
projects suggested in Statistics By Example as exercises (e.
g., making up a small, but real, food price index). Then experiences and suggestions might be put together in a fresh
publication, Statistics by Project.

Reaching out. The influence of these new teaching materials
in statistics should not be restricted to teachers whose
field is primarily mathematics. Important statistical issues,
ideas, and methods arise in the laboratory sciences, the social
sciences, even in language studies and the arts. The Guide's
essays are organized by general field, and the segments of
Statistics by Example are readily so classified.

There are opportunities through links with teachers' socie-
ties, through encouragement of individual teachers in the
schools, through writings and talks, to bring an improved
measure of rapprochement. A particularly important field -
mentioned earlier - is the introductory laboratory course.

Another mode of reaching out was suggested to us by Richard
Savage. We call it the statistical field trip concept.

I don't know how frequently field trips occur in the schools
of other countries; in many schools in the U.S.A. they are
fairly common. Usually they will, at the secondary school
level, be to museums, concerts, or the like. Less often they
will be to factories, legislative bodies, courts, and so on.

Savage's idea was to build on the wide use of statistical
ideas and methods in modern society. For example, a class
might visit a factory and have its quality control program
explained and exhibited: control charts, inspection methods,
corrective measures, perhaps acceptance sampling. There might
be a visit to a municipal water testing laboratory, or an air
pollution measurement center. Other possibilities are fire and
police departments, banks, industrial or agricultural testing
facilities. Possibly even gambling casinos or race tracks -
naturally in their off hours - might be willing to take stu -
dents on an explanatory tour. (After all they will soon be
adults.)

So far this is mainly a idea, to my knowledge. I should like
to see some trials, with careful, detailed preparations. For
one thing, it may not be easy to find guides at the establishments visited with suitable understanding of statistics.

The Committee may encourage this kind of trial and later report
on what happened and what lessons may be learned for future
trials.

Teacher training.

The most frequent reaction we have had from secondary school
teachers, and others who know American secondary schools, is

that the Committees publications will not be used as effectively and widely as they might be without a sustained effort
to provide background and understanding for teachers. We have,
of course, teachers' manuals for Statistics By Example, yet
it is felt that these will not be enough, that direct training will often be required.

There are many ways in which such training might be arranged,
and the Committee will doubtless try for trials of such ways.
Detailed plans and the financial underpinnings, have not been
fully discussed at the date this summary is written.

Experiments

Statisticians, like doctors, preachers, and other human beings,
frequently do not follow their own professional precepts. In
particular, only rarely do statisticians suggest proper ex -
periments for evaluating new initiatives in teaching probability and statistics. Here I should emphasize the familiar dis -
tinction between innovative trials, on the one hand, and true
experiments on the other. Innoviate trials are frequent in the
schools; although they may be suggestive and helpful, they are
often wholly inconclusive, non-cumulative, and with limited
persuasive force. Proper experiments, on the other hand, with
appropriate control groups and randomization, are more likely
to contribute to the cumulative growth of knowledge, more likely to be persuasive, and less likely to mislead because of
bias or non-quantified adventitious occurrences.

As we all know, it is difficult to do careful experiments in
education. First, there is the Hawthorne effect --- student
and teacher alike may behave differently when they know themselves to be in an experimental situation. Second, there may
be a novelty effect; the excitement of something new may by
itself promote special interest and industry that would not
last if the something new becomes routine. (Indeed, that is
sometimes a good argument for trying out new methods simply
for the sake of novelty.) Third, there is the problem of criteria. Fourth, it is often difficult to know what to treat as

basic experimental units (students? classrooms? schools? cities?) and to keep their behaviour substantially independent. No doubt I've left out a number of other serious difficulties. Nonetheless, experiments - not perfect, but good - ~~can~~ be done in education. If we propose

new paths in statistical education, it behooves us as statisticians to consider evaluation of those paths, and the very best evaluation is via experiment.

Another suggestion along this line is the carrying out of modest local experiments within a school on a variety of questions, with design, execution, and analysis done as a student-teacher project. One can think of all sorts of operational areas for this, from lunch-room queueing to room scheduling, from training methods in athletics to fire-drill frequency or timing, from teaching spelling (where it is relatively easy to set criteria) to effects of articles in the school news - paper on student opinion. Even if the experiments are imperfectly done - and they frequently will be - exposure to the experimental method, and learning by making mistakes, may be helpful.

Final remarks

I have attempted to set out in organized fashion the possible future activities that the Joint Committee may point towards. It clearly will not undertake all these activities, and it may move in unanticipated directions. The Committee has a keen interest in the content of our discussion here. This select, international group with wide experiences is likely to bring fresh breath to our committee deliberations.

Which possible activities are especially appropriate for international discussion? Are some so special to the cultural and educational frameworks of particular countries that international discussion would be fruitless? An example might be the statistical field trip concept, where the details - indeed the whole framework - might differ substantially among the countries. Even then, of course, international discussion may in principle be helpful.

The cross-cultural point is an interesting one. Will readers from other lands of Martha Zelinka's baseball example be led to think of analogous examples from cricket, football, or even chess?

Let me also ask you which of these possible activities are most appropriate for a sustained International Statistical Institute contribution. Which are appropriate for other kinds of organizations, and which for private efforts?

The members of these Committee look forward to learning the advice and recommendations that will flow from this meeting in Vienna.

DISCUSSION

J. Oyelese: Professor Kruskal, in his paper, has suggested the statistical field trip concept. The idea of a field trip or excursion by secondary school pupils is fairly common in Nigerian schools. These trips are usually organized by school clubs such as the Geographical Society, the Historical Society and the Science Society. The trips are usually made to places of interest to the subject area. There have recently been trips made by secondary school pupils under the direction of the class teacher to see the computer at the University of Ibadan (and other places).

C.R.Rao: Field trips for students taking courses in statistics serve many important purposes. a) A visit to a factory using statistical quality control methods will enable them to see how the techniques learnt in a class room are used in practice and what benefits result from using them. b) Field trips for collecting data will help the students in learning the different aspects of sample survey. c) Field trips are generally liked by students. They can be used for educative purposes, more valuable than classroom lectures, provided they are properly organized and led by experienced professors.

H. Aigner: Types of field trips practised in Austria: a) Half-day or full-day trips supplementing physical training (hiking trips primarily; references to other disciplines are marginal and depend on the accompaniyng teacher, who is not usually a physical - training teacher.

b) Half-day or full-day trips supplementing various disciplines (trips to museums, industrial plants, governmental bodies,
airports, etc. for a given purpose), c) Trips of typically a
week's duration supplementing physical training (ski-trips,
scant references to other disciplines), d) Trips of typically
a week's duration supplementing social education (experiment
in rural living, often combined with science: botany, sur -
veying, etc.), e) Trips of typically a week's duration afford-
ing an overview of industrial (economic) practice in-the
student's field (often including a portion abroad). Often the
trips have an unplanned placement function.

A. Engel: In Germany a class has usually each year one field
trip of one day duration and several of 1/2 day. In the one
day trip students must walk at least 20 kilometers (about 12
miles). So the class is mostly driven by bus far away from
home, then they walk to some destination, for instance a
brewery or salt mine or what not, which is inspected. Mostly
an engineer gives details of the operation of the company.
More interesting is the so-called "Schullandheim". It occurs
only twice in the whole school time. For instance in grade 8
the whole class goes to the country side and stays for two
weeks near a small community. They have about 3 hours of instruction per day. The remaining time is taken up by projects.
The class is devided into 2 groups and more. Each group gets
a special assignment. The aim of the class is a detailed investigation of the community: the geography, botany, zoology;
the political, economical, social structure. The results are
written up as a book. Each group contributes one chapter to
the book. Before going to the countryside the preparations
begin at least 6 weeks in advance.

L. Råde: One project I would like to recommend for the ISI
educational committee and the JCCSP committee is the writing
of a handbook in teaching probability statistics at the school
level. Such a handbook should be very valuable for countries
starting experiments in the teaching of probability and statistics at the school level. Also for individual teachers such
a handbook would be of great help. Such a handbook would
contain e.g. the following things, 1. Didactical analysis of
different approaches to specific parts of a course in probability and statistics, 2. Essays on e.g. the use of graphical
methods, use of experiments, use of computers and so on, 3.
A selected bibliography with short summaries of papers and
books mentioned. 4. Information about experiments undertaken
in the field of teaching probability and statistics at the
school level. 5. Problems classified according to different
categories (field of application, mathematical difficulty and
so on.)

H. Aigner: Teachers would welcome a reference book on the
lines of the German "Handbuch der Schulmathematik", but avoiding two of its inherent flaws: a) The proposed handbook should
be in loose-leaf format, so that new ideas can be added prompt-
ly (without waiting years for a new edition), b) It should be
published in several major languages (also with "cultural"
modifications (sports, currency) etc. for different countries.

TEACHING OF STATISTICS AT THE SECONDARY LEVEL

T. POSTELNICU

Academia Republicii Socialiste Romania, Bucarest

Since 1968, in the Socialist Repulic of Romania, probability and statistics have been taught in the last school year (the 12th one).

A course for the students, entitled "Elements of Probability and Statistics", has been published under the guidance of Professor Gheorghe Milhoc, Director of the Centre of Mathematical Statistics of the Academy of Romania. This course has six chapters from which the following four are included in the curriculum:

1. Finite probability spaces
2. Classical probabilistic formulae
3. Random variables. Expectation and variance
4. Elements of mathematical statistics.

This last chapter contains the following topics:

- Basic notions of mathematical statistics
- Graphical representation of statistical data
- Characteristics of statistical data
- Theoretical and empirical distributions

Various numerical examples and 15-20 problems are added to each chapter.

The other two chapters of the course are:

5. Characteristic values for classical probabilistic schemes
6. Infinite probability spaces

These topics are to be taught also in the last year of school, but only in the special classes with mathematical profile.

tribution.

It is my opinion that this amount of probability and statistics is at present the maximum we can cover in our secondary schools. It seems important to me that the student at least get to know the elements of statistics and that they can understand graphical representations of statistical data.

Finally, I want to list some questions.

1. How much statistics should be taught in secondary schools?
2. How much statistical theory should one treat in secondary schools?
3. How is an interdisciplinary approach, relating mathematics and statistics to physics, biology, geography and economics, achieved?
4. How much statistics should be treated in commercial and technical schools?
5. How can the training of secondary school teachers in probability and statistics be improved? (In Austria, as far as I know, no course in statistics is compulsory for teachers examinations.)

INTRODUCTION OF STATISTICS IN THE SECONDARY SCHOOL

I. P. RUCKER

State Department of Education Richmond, Virginia

Background

The data used in this paper were secured directly from certain high school classroom teachers who were, in 1972-73 , teaching probability and/or statistics in the schools of the United States and some of its possessions. The names of the teachers were provided to the writer of this paper by the State Supervisors of Mathematics throughout the country. 69 names were submitted. Each of these 69 individuals was sent a questionnaire and asked to respond to it. (A copy of the questionnaire is found in Appendix A.) 50 of the 69 teachers returned the completed forms.

Does the above imply that only 69 teachers in the United States are teaching probability and/or statistics? No! In the first place, the number of names from each state was arbitrarily limited (by the writer)to five. Secondly, not all states provided names; some, because none of the schools in their states was on record as teaching probability and/or statistics as a course during 1972-73; others, a small number, did not respond.

This paper will not attempt to separate probability from statistics because of the interrelated nature of the two and because, in the high schools of the United States, the two, if considered at all, are generally considered together. Six aspects of introducing probability and statistics in the secondary school will be presented: the teacher, the student, the course, the school, the impetus from the state education agency level in two states, and some specific examples of successfully introduced courses.

The Teacher

1. What courses should a teacher of probability and statistics have taken prior to teaching the course?

   The response to this question varied; however, the ma - jority of the teachers felt that the following courses should be required: undergraduate level probability and statistics including introduction to probability and statistical inference, computer mathematics, statistical applications, analysis of variance, game theory, linear programming, logic, and hypothetical testing.

   Other courses recommended are: number theory, theoretical statistics, advanced probability and statistics, and mathematical statistics.

2. Are courses in probability and statistics generally available to undergraduate students?

   In general, courses in probability and statistics are available in the colleges and universities. The problem is that most institutions of higher learning do not require these courses of their mathematics major. Since this is true, many of those who wish to teach mathematics either do not know about the courses or they prefer to take other courses.

3. If a teacher is not prepared to teach probability and statistics prior to becoming a teacher, how can he prepare himself?

   a) About half of the reporting teachers had probability and statistics in their undergraduate and graduate training. b) The others noted that they had learned the subject on their own and/or along with their students. The latter procedure requires many hours of work on the part of the teacher and often leaves him without the capability of responding to many of the searching questions which students pose. c) Some of the teachers sought out and found summer institutes in which they could take probability and statistics, while others took Saturday courses at nearby colleges.

4. Why do some teachers want to teach probability and statistics?

   a) Most of the teachers said that they had a fascination for the subject, a general interest in mathematics with specific interest in its applications to current events , and an interest in and responsiveness to students' needs. b) One teacher noted, quite honestly, that since only the better students take the course, the classes are small and that this in itself can be an incentive !

5. Does the teaching of probability and statistics take more teacher-preparation time than other courses?

27 of the teachers reported that it does; 23 that it does not. The amount of preparation time depends largely upon the background of the teachers; therefore, the more indepth knowledge of probability and statistics, the less preparation time is required. No teachers responding to the questionnaire said that the extra work involved in preparation for classes was a burden; 39 said that it was a pleasant task; 11, that it was "just a part of the job"

6. What other courses do teachers of probability and statistics often teach ?

For the most part, teachers of probability and statistics teach the higher level academic mathematics courses. Several reported that they teach drafting, general mathematics, consumer mathematics, and pre-algebra courses.

7. If a teacher of probability and statistics were asked to relinquish one of the courses they teach, which one would they wish to give up?

Those teachers teaching the low-level classes (general mathematics, pre-algebra, etc.) were in accord that they would like to relinquish these. The majority of those who teach geometry reported that they would willingly give up this course. Three of the teachers indicated that they would like to relinquish the probability and statistics course. Their reasons were not that they dislike the course or the subject matter, or the students, but that because of a) a heavy teaching load, there was not enough time available to prepare adequately for the probability and statistics course; b) insufficient background in probability and statistics to feel comfortable with it; and, c) a preference for teaching analysis, calculus, and other upper level mathematics courses.

The Student

1. Do more boys or girls take probability and statistics?

According to the teachers reporting, more boys than girls take the course. Only two reported that there were more girls than boys in the classes.

2. Has student interest in probability and statistics in - creased or decreased during the past five years?

60% of the teachers reported that student interest is on the rise. This is attributed partly to increased school enrollment, but largely to students' awareness of the uti-

litarian aspect of probability and statistics in many
fields of endeavor - including everyday living.

3. Why do students take the course?

a) Students who elect probability and statistics are usually not only the capable students, but those who have a
curiosity to learn something about the subject. They are
inherently motivated. b) Many non-scientific students recognize the applicability of probability and statistics
to field other than the sciences. c) Other students take
it "just for fun". d) One teacher reported that some of
his students take it "to avoid study hall"!

4. What is the attitude of the students toward the course?

At the outset of the course, it appears that most of the
students have a healthy, but sometimes neutral, attitude
toward the course. As the course progresses, this turns
significantly upward to a positive attitude. Only two teachers reported that they sensed the attitude of their stu -
dents to change from positive to negative.

5. Do many students drop the course before completing it?

No! Only three teachers reported that the drop-out rate
was high. The reasons for any drop-outs, generally, is not
lack of interest; rather, it is that the students who drop
out, are carrying a heavy load of required courses, and
they find that time to study probability and statistics is
not available to them.

6. Are the follow- up studies made of students after they leave
high school?

Generally the answer was "no". Six teachers have followed
through and have been pleased to learn that several of the
students were so inspired by their introduction to probability and statistics in high school that they are majoring
in it in college. The number of such students is small, but
worthy of mention especially since probability and statistics is relatively new to the high school curriculum. Some
students have provided feed-back to their teachers which
has helped in improving the high school course.

The Course

1. How long has probability and statistics been offered in
the high school?

Prior to the decade of the 1960's, it was a rarity for a
high school to offer probability and statistics as a separate course. In some of the high school mathematics texts,

there was a brief mention of the subject and some few
exercises that could be treated as experiments. In too
many instances, even these were omitted. The immediate
reason for this was given as, "We don't have enough
time for everything, so we omit the frills". The real
and understandable reason was that many teachers were
unprepared to cope with the content. This statement is
not to be interpreted as an incrimination of teachers
but as an indication of the trend of those times to
overlook courses in probability and statistics.

Due primarily to the impact of the 1958 report of the
Commission on Mathematics of the College Entrance Examination Board, the course has begun to filter into
the schools. The following data attest to this fact:
Six teachers reported that the course has been offered
in their school for one year; three for two years; nine
for three years; six for four years; five for five years,
four for six years, one for eight years; four for ten
years; one for eleven years; one for twelve years; and
one for fifteen years. (As a point of interest, the person who reported that the course has been offered for
fifteen years was a member of the Commission on Mathematics!
He teaches in an independent school and began the course
on a trial basis, in 1958, until administrators and fa -
culty were convinced that the course is a worthy one for
high school students.)

2. How can a course in probability and statistics be introduced?

   The teachers reported in the following ways that the
   course was inaugurated in their various schools: creating
   interest on the part of the scientific as well as the non-
   scientific (particularly social sciences) students; arous-
   ing interest on the part of teachers to tackle the job and
   to prepare themselves to do so; teachers' willingness to
   take on an extra class (probability and statistics) on a
   pilot basis, then make it work; a strong consideration for
   including probability and statistics when expansion of the
   offering is anticipated; giving evidence to the administrative staff that the course has practical as well as theoretical applications.

3. What are some of the problems involved in introducing probability and statistics into the secondary school?

   a) The major problem is the lack of adequately trained personnel to teach the course. As indicated earlier in this
   paper, many of the teachers surveyed had learned probability and statistics either on their own or along with the
   students, or both. b) The problem of convincing administrators often looms heavily. c) Scheduling was noted as a problem because most high school schedules are already very
   tight. d) Ways to motivate students to take the course was

cited by some teachers. e) Ways to motivate teachers to prepare themselves to teach the course is a gnawing problem compounded with the lamentable fact that probability and statistics courses which are designed for teachers are often non-existent.

4. How can the problem be solved?

a) Identifying at least one teacher who is willing to prepare himself, tackle the course, and (if necessary) offer his free time to work with a few students at the start. From this point on, if the course is well-taught, the students will "sell" the course to others. Once students, in sufficient number to offer the course in the regular schedule request it, the administrator becomes convinced, the other teachers catch the desire teach the course and prepare themselves, and the course is introduced on a regular basis. b) Scheduling problems solve themselves once the procedure in 4 (a) is followed. c) One way to motivate students is to incorporate units on probability and/or statistics in established courses. One teacher reported that he teaches a course in computer mathematics. He saw the opportunity to combine statistics with this course and proceeded to do so. In time, the introduction of a full-blown statistics course became a reality. d) Not all administrators have to be convinced. One such person decided that the course should be a part of the regular offering; hence, he assigned a teacher to teach it, provided means by which that teacher could get the necessary training, and included the course in the scheduling. This worked! (Caution in the use of such practice is noted. The situation, especially including the rapport that the administrator has with the staff, must be ideal.)

5. Is the course elective or required ?

In all instances, the course is elective on the part of the students.

6. What prerequisites must students meet?

Although some schools accept students who have had only a basic course in elementary algebra, most of the schools require at least two years of algebra and one of geometry.

7. What is the duration of the course?

30 of the teachers reported that their courses run for one semester; 20 reported two semesters. For those schools that are either on or are contemplating a trimester plan, it is strongly recommended that the probability course extend over at least two of the three trimesters.

8. Is the course a laboratory course?

A small percentage of the teachers reported that their course is laboratory oriented. Most of the teachers, 39, said that theirs is not. Many schools do provide desk calculators but beyond that, the equipment and materials for scientific investigation of statistical ideas are lacking.

9. What is the cost of equipment for probability and statistics ?

The cost range reported is from 50 to 12,000 dollar.

10. What is the name of the course as offered in the schools?

An overwhelming majority of schools call the course "Probability and Statistics". Other names given are: " Pro - bability"; "Statistics"; "Statistics and Finance"; "Statistics and Society"; "Introduction to Probability and Sta - tistical Inference"; and, "Advanced Mathematics IV".

11. Are the texts that are currently on the market suitable for high school students?

The teachers felt that while the number of texts in probability and statistics is small, those that are available are excellent. The four (4) texts mentioned, in order of frequency of use, are:

a) Mosteller, et al, Probability: A First Course, Addison Wesley Publishing Company.

b) Mosteller, et al, Probability with Statistical Applications, Addison Wesley Publishing Company.

c) Willoughby, Stephen, Probability and Statistics, Silver Burdett Publishing Company.

d) Blakesley, et al, Introduction to Statistics and Probability, Houghton Mifflin Publishing Company.

The School

1. Is the course generally available in rural, suburban, or urban schools?

Four teachers reported that their schools were in rural areas; twenty, in suburban areas; and, twenty-two in urban areas.

2. What is the student enrolment of a school where proba -
   bility and statistics is offered?

   Four of the teachers are in a school with enrolment less
   than 500; eight, with 501-1000; thirteen, with 1001-1500;
   eleven, with 1501-2000; thirteen with over 2000 enrolment.

3. What grade levels do the schools include?

   Twenty-two reported grades 9-12; twenty-two, grades 10-12;
   one, grades 11-12.

The Impetus from Two State Education Agencies

1. NORTH CAROLINA

The following quote is from the Director of the Division of
Mathematics, Department of Public Instruction, Raleigh,
North Carolina:

".... We are encouraging schools to offer Probability and
Statistics as a short course, or mini-course, for senior
level students. A number of schools are interested in mini-
courses in mathematics at this time, so we have prepared a
series of performance objectives for such a course in Probability and Statistics. ...We hope that more and more schools
will move to some type of offering in Probability and Statistics within the next couple of years. In fact, many of our
college professors recommend Probability and Statistics along
with several other topics to be taught in lieu of a full
year's course in Calcus. I believe that this is the direc -
tion most of our schools are trying to take as we move toward
a five-year program for College-bound students."

The student performance objectives for probability and statistics which have been prepared for the State of North Carolina
are found in the Appendix B.

TEXAS

The program Director of Mathematics, Texas Education Agency,
says:" Although we have used Probability and Statistics as

an approved course since 1961, there are not many takers.
Interest seems to be growing, however, with increased consumer
needs for statistical know-how.

"Our new mathematics bulletin describes the course which I
enclose. (See Appendix C.) It looks elementary, but acquires
stature and rigor with the teachers and with a book such as
Mosteller."

Specific Examples of Successfully Introduced Courses

In 1970, the writer surveyed the local supervisors of mathematics throughout her state (Virginia) to ascertain the kinds
of current activities provided for high school students interested in probability and statistics. The following listing is
gleaned from that survey.

a) A two-semester course called "Computers and Statistics"
is offered. Students learn computer programming so that they
can use the computer as a tool in solving statistical problems.

b) Statistics as a one-semester course is an adjunct to a
one-semester course in analytic geometry.

c) In a course called "Advanced Algebra", 50% of the content
is probability.

d) A program called "The Able and Ambitious Summer Program"
is designed for rising junior and senior high school students
whose abilities, industry, and interests are at a high level.
The first summer of its inception, fifteen (15) students elected mathematics and, in particular, probability and statistics. No specific text was used; rather, a variety of topics
was explored. One of the interesting experiments carried out
was that of observing the flow of traffic on a specified down-
town street during a specified time-period of one day. The
students' observations prompted them to invite the traffic
director for a day's consultation. The students explained to

this director which traffic lights were off-timing, how traffic could be more quickly and efficiently funneled through the street, and the results of other observations that would expedite movement of traffic along that particular street. The traffic engineer was impressed and set in motion the machinery to improve the situation !

e) All senior high schools in one large school division are equipped with desk-top computer calculators. In addition, computer terminals have been installed in all high schools and are used to enrich and extend the two-semester course in probability and statistics. In this course, students are required to conduct independent research studies and to sub - mit statistical analysis for their findings. These studies are, most often, non-trivial in nature because of the students' accessibility to the computer.

Of Recent Development

During the summer of 1973, the first Governor's School for the Gifted was held in Virginia. The school accomodated 400 gifted and/or talented rising juniors from the public and private high schools of the State. One of the offerings during the School was Computers and Statistics with Probability. The course was designed, of course, as a pre-calculus one. The students applied their knowledge of mathematics (which was considerable for the general run of high school students) to the problems of statistics and probability (about which few knew much, if anything) and used the computer for problem solving. Enthusiasm was so high among those who took advantage of the course that we are working toward a follow - up of these students during the ensuing school year. We hope to provide means by which the students can continue their interest in their own high schools where neither computers nor probability and statistics is available.

The impact that these students will have on getting probability and statistics introduced in their schools is uncertain

at this time. With the support of the state education
agency, perhaps something far-reaching in this area will
evolve. We shall try !

APPENDIX A

PROBABILITY AND STATISTICS

A Questionnaire

A. The Teacher

1. Approximately how many years have you been teaching?
   1-5____ 6-10____ 11-15____ 16-20____ More than 20____

2. Are you currently assigned to the mathematics department?
   Yes____ No____.
   Are you head of the department? Yes____ No____.

3. How long have you been teaching probability and/or sta -
   tistics? _____

4. What prompted you to become interested in teaching probability and statistics ?
   _____

5. Did you learn probability and statistics
   a. in undergraduate school? _____
   b. in graduate school _____
   c. on your own? _____
   d. along with your students? _____

6. What courses do you feel that a teacher should have taken
   in order to do the best job of teaching probability and
   statistics? _____
   _____

7. Do you feel that these courses are generally available to
   interested undergraduate students? Yes____ No____

8. Are you the only one teaching probability and/or statistics
   in your school? Yes____ No____. If "No", how many others
   are there? _____

9. How many courses, other than probability and statistics,
   do you teach? _____

What are they? _____

_____

10. Do you feel that you must do more preparation for the
    probability and statistics course than for other cour-
    ses that you teach? Yes____ No____.
    If "Yes", is this extra work a burden? ____ a pleasure?
    _____ "just a part of the job?" ____

11. If you were asked to relinquish one of your courses, which
    one would you like to give up? _____

12. If "probability and statistics" is your answer for ques -
    tion 11, please explain. _____

    _____

B. The Student

1. Is student interest in taking the course high? ____ low?

   _____.

2. Why do students take the course? _____

   _____

3. Does student attitude toward probability and statistics
   change as the course progresses? Yes____ No____
   If "Yes", is the change positive to negative? ____ negative to positive?_____

4. Are there more boys____ or more girls _____ taking probability and statistics in your school?

5. What is the approximate size of your class? _____

6. Has this number increased____ or decreased _____ over
   the past five years?

7. Is the dropout rate in probability and statistics high?
   _____ or low?_____

8. To what do you attribute the dropout rate? (Don't be
   modest!)_____

   _____

9. Are most of the probability and statistics students college bound? Yes____ No____

10. Have you made follow-up studies of your students after they leave high school? Yes____ No____
If "Yes", would you please share some of your findings?

_____

_____

C. The Course

1. How long has probability and statistics been offered in your school? _____

2. What prompted the inclusion of probability and statis - tics in your curriculum? _____

3. What were some of the problems (assuming that there were some!) in getting the course inaugurated? (It would help to know, for example, if the problems were administra - tive, financial, scheduling, and/or lack of qualified personnel to teach the course.) _____

_____

_____

4. How were the problems noted in Question 3 solved?_____

_____

5. What is the name of the course in probability and statis- tics?_____

6. Is it one semester?____ two semesters?____ other? (Please explain.) _____

7. What is the name of the textbook that you are currently using?
Author:_____
Publisher:_____
Title:_____

8. Is there a laboratory to accompany the course? Yes___No____

9. If the answer to Question 8 is "Yes", please indicate in
   the following:
   a. amount of time per week (in hours)_____
   b. approximate cost of the equipment _____
   c. sources from which the equipment was provided:
      school district____ school____ other (Please explain)

   _____

10. For what grade levels is the course designed? 10___ 11___
    12___

11. What are the prerequisites for the course? _____

    _____

12. Is the course elective?____ or required?____. If required,
    is this for all or certain students? (Please explain.)___

    _____

13. Have you used the kinescopes, "Probability and Statistics",
    which were made from the Continental Classroom television
    series? Yes____ No____. Don't know about them_____

D. The School

1. Is your school rural?____ suburban?____ urban?

2. What grade levels does your school house? _____

3. What is the approximate enrollment of your school?
   Less than 500____ 501-1000____ 1001-1500____ 1501-2,000
   ____ Over 2000____

E. Other Comments: (Use back of sheet, if necessary.) _____

   _____
   _____
   _____

I am ____ am not ____ interested in receiving, after September 15, 1973, a copy of the paper to be presented at the biennial meeting of the International Statistical Institute.

Name_____ Mailing Address _____

APPENDIX B

PROBABILITY AND STATISTICS - PERFORMANCE OBJECTIVES

On completion of the course, the student should be able to:

1. interpret data presented in tabular and graphical form.

2. organize and present data in tabular and graphical form.
   For example: bar graph, broken line graph, curve line
   graph, circle graph, pictorial chart, histogram, frequency polygon, and cumulative frequency polygon.

3. summarize and analyze selected data by calculating the
   measure of central tendency (mean, median, mode); discuss their relative merits.

4. calculate measures of dispersion: range, quartile deviation, variance, and standard deviation.

5. compute measures of central tendency and measures of dispersion from ungrouped data.

6. compute the mean and standard deviation for a large number of measurements by grouping techniques.

7. determine whether outcomes of an experiment are equally
   likely.

8. determine whether outcomes of an experiment are mutually
   exclusive.

9. determine the sample spaces of various experiments

10. express events as subsets of a sample space.

11. determine whether two events are complementary.

12. determine whether events are exhaustive.

13. develop and apply the formulas for calculating the num-
    ber of permutations and combinations of n objects taken
    r at a time, and determine whether the problem involves
    permutation or combination.

14. apply correctly the two basic counting principles to determine the number of outcomes in event "A or B" and event "A and B".

15. determine the probability of an event in a finite sample space using the classical definition of theoretical probability.

16. estimate the probability of an event by the relative frequency concept in a series of experiments.

17. apply combinatorial theory to calculate the probability of events.

18. identify binomial experiments and apply the laws of chance to the binomial distribution.

19. compute the conditional probability of event A given B.

20. determine whether two events are independent.

21. relate the probability of occurrence of two independent events to the probability of their intersection and apply to appropriate problems.

22. compute the mathematical expectation of an event from an actual or theoretical experiment.

23. apply the appropriate additive and multiplicative theorems to determine the probability of multiple events.

24. distinguish between samples and populations and between statistics and parameters.

25. use elementary sampling theory to perform a random sampling.

26. apply the theory of probability to acceptance sampling.

27. apply the theory of probability to test statistical hypotheses involving normal and binomial distributions.

28. use samples to make estimates of population measures.

APPENDIX C

PROBABILITY AND STATISTICS

The increasing significance of probability and statistics in our society makes it necessary to offer an opportunity for students to gain knowledge in these areas. This course is designed to give the student a one-semester, pre-calculus experience in probability and statistics.

COURSE DEVELOPMENT

I.   Permutations, combinations, and the binomial theorem
     A. Permutations
         1. Tree diagrams
         2. Multiplication principle
         3. Factorial symbolism
         4. Formulas
         5. Permutations involving two or more like items
         6. Circular permutations

     B. Combinations
         1. Difference in permutations and combinations
         2. Formulas

     C. Binomial theorem
         1. Binomial expansion using combinations
         2. Proof by induction

II. Equally likely outcomes
     A. Experiments
     B. Sample space
     C. Probability
     D. Events
         1. Mutually exclusive
         2. Independent
     E. Conditional probability
     F. Random numbers
     G. Product rule
     H. Bayes' theorem

III.   Random variables
    A.   Probability function
    B.   Mean
        1.   Arithmetic
        2.   Geometric
    C.   Average and standard deviation
    D.   Chebyshev's theorems

IV.   Joint and continuous distributions
    A.   Probabilities represented by areas
    B.   Cumulative probability graphs
    C.   Normal curve

V.   The binomial distribution
    A.   Binomial experiments
    B.   Binomial probability tables

VI.   Inferential statistics
    A.   Theory of sampling
    B.   Hypothesis testing

DISCUSSIONS

C. R. Rao: Information collected by Mrs. Rucker is very valuable. In the light of what she said we have to consider two
important problems: a) The content of a Statistics Course at
the High School level, methods of teaching and preparation of
course material, b) Training of teachers to give courses in
statistics.

B. Benjamin: In the U.K. most people teaching "statistics and
probability" would not feel disposed to mention "probability"
separately; They would simply speak of "statistics" and assume
that it would be understood that "probability" was included.
There would many teachers making brief references to probability in the course of teaching more advanced algebra but they
would hardly regard themselves as teaching probability, and
would not give a positive answer to Dr. Rucker. Because so much
probability is taught in order to test the agreement of expectation and observation, we would rarely agree to separate probability from statistics. Further more, in the U.K. most ele-

mentary courses in statistics teach the concepts of probability well on into the course after frequency distribution, measures of central tendency and dispersion, and correlation, but <u>before</u> sampling and estimation. For this reason we tend to use textbooks which do <u>not</u> specifically mention probability in their title. Books on probability might be used in addition but not frequently.

A SURVEY OF THE TEACHING OF STATISTICS IN SECONDARY GRAMMAR SCHOOLS IN THE WESTERN AND LAGOS STATES OF NIGERIA

J. O. OYELESE

University of Ibadan, Ibadan

The purpose of this report is to describe the present position of the teaching of Statistics in two of Nigeria's twelve
states. The situation in the other states of Nigeria as well
as in most English speaking countries in West Africa is very
similar to the one described here.

Nigeria is a federation of twelve states, diverse in size ,
population, economic, social, cultural and educational development. The Western and Lagos States of the Federation covered by this survey has an area of 34,847 square miles (a little
less than one-tenth of the total area of the country) and a
combined population of 10,921,093 out of a total population
of 55,620,2681 (1963 Census) for the whole country.

In most parts of the Federation primary schools offer a six-
year course while secondary grammar schools offer a five-year
course leading to the West African School Certificate Examination. This is followed, in a selected number of schools, by
a two-year Higher School Certificate Course to provide further
education for the more able students.

For the purpose of this survey a questionnaire seeking information on the teaching of Statistics was sent to all the 37
schools in the Lagos State and all the 194 schools in the
Western State recognised in 1970 for the West African School
Certificate and Higher School Certificate Examinations by
the West African Examinations Council. The total of 231
accounted for about one-third of all the 670 schools so recognised all over the Federation by the West African Examinations Council. There are of course many more secondary schools

in the Federation and in the two states than the number given above would indicate; this is so because only grammar schools which have existed for five or more years are recognised by the West African Examinations Council for the School Certificate Examination. Table 1 below gives the number of secondary schools of all types in the Federation as a whole as well as in the two states covered by this survey.

TABLE 1

Secondary Education: Number of Schools, Teachers and Pupils by Sex, 1970 1)

No.of Schools	Teachers			Pupils		
	M	F	MF	M	F	MF
Lagos 87	873	380	1253	15,487	11,725	27,213
West 477	3,959	1,029	4,988	70,404	41,598	112,002
Nigeria 1,155	11,424	2,667	14,091	205,959	104,095	310,054

1 Source: Annual Abstract of Statistics, Nigeria 1971
         Federal Office of Statistics

It will be seen that 49 per cent of secondary schools of all types and about 45 per cent of the school population in secondary schools are from the two states.

Only 85 (about 37 per cent) of the schools returned the questionnaires, some of them after two reminders. Of this number 60 gave a positive response indicating that Additional Mathematics, which includes Statistics, was taught, while the remaining 25 schools did not offer the subject.

Elementary Mathematics is compulsory in all secondary grammar
schools in Nigeria and until recently the majority of all
schools still offered only traditional mathematics. Traditional Additional Mathematics, which is an optional subject, is
taught only in schools where there is an adequate mathematics
staff and where sufficient interest in mathematics has been
generated among the pupils. Examination in each of two papers
in the subject is made up of two parts: Section A - which consists of five simple compulsory questions in elementary calculus, trigonometry, algebra and statistics. Section B has
six slightly more difficult questions in Pure Mathematics (3),
Statistics (2) and Mechanics (1). It appears possible to pass
creditably in the examination without answering any of the
questions in Statistics and this is what some pupils probably
do. Some teachers who responded to the questionnaire thought
that the amount of time spent on Statistics in Section A is
not justified considering that only one question is set on
the topic (see Appendix for the Statistics Section of the
Traditional Additional Mathematics Syllabus).

Statistics is offered in four or five different forms in the
secondary grammar school. First, it is taken as one of the
topics in Additional Traditional Mathematics which has been
described above. Secondly, Statistics is now a common feature
of every modern mathematics syllabus. The new Elementary
Mathematics Alternative "C" syllabus has a small section on
Statistics (see Appendix for the scope of this). Thirdly, the
Alternative Modern Additional Mathematics of both the Entebbe
and the Joint Schools Project Courses have sections on Statistics. The two different courses have now been harmonised into
a new Modern Additional Mathematics. Fourthly, there is a
small section on Statistics in the General Paper of the Higher
School Certificate Examination and the mathematics syllabus of
the Advanced Level General Certificate of Education and Higher
School Certificate Examinations. The content of this syllabus
also appears in the Appendix.

There has been a steady swing towards the introduction of modern mathematics in a good number of schools. The only handicap seems to be the lack of qualified teachers and the difficulty experienced by many schools in getting suitable texts.

Table 2 below gives the number of teachers in secondary grammar schools in the two states compared with the rest of the country. This table shows quite clearly that non-graduate teachers are in the majority. Some of the schools responding to the questionnaire had only teachers with either a school certificate or a Higher School Certificate in charge of the Additional Mathematics which includes Statistics.

TABLE 2

Qualifications of Teachers in Secondary Grammar Schools[2]

GRADUATES			NON-GRADUATES			
With t. qu.	without t. qu.	NCE	Grade I or equivalent	Grade II	Other	Total
Lagos						
M 133	M 130	M 85	M 59	M 24	M 127	
F 75	F 66	F 75	F 29	F 81	F 33	844
West						
M 515	M 762	M 491	M 175	M 56	M 972	
F 88	F 204	F 120	F 21	F 17	F 304	3,725
Nigeria						
M 1,966	M 1,744	M 1,389	M 526	M 696	M 3,326	
F 450	F 467	F 401	F 93	F 193	F 635	11,917

[2] Source: Annual Abstract of Statistics, Nigeria 1971
Federal Office of Statistics

Notes

t. qu.   = teaching qualifications

NCE   = National Certificate in Education. This qualification is obtained by spending three years in a
college of Education at a University or one that
is affiliated to a University.

Grade I   = A trained teacher with either the Higher School
Certificate or the General Certificate of Education at the Advanced Level in two teaching subjects.

Grade II   = A teacher who has undergone a three year training programme at a Teacher Training College receives this qualification at the end of his
training.

Other   = may include teachers with only the West African
School Certificate or the General Certificate of
Education at the ordinary Level.

The staffing position is a little more acute in Girls' Secondary Grammar Schools. Of the ten schools recognised in the
Lagos State for the West African School Certificate Examination only four offer Traditional Additional Mathematics. The
situation is not any better in the Western State where only
four of twenty-five or so recognised schools teach statistics.
The difficulty in recruiting mathematics staff may be somewhat eased in the near future with the development of the recently launched National Youth Service Corps in which graduates are posted all over the Federation - mainly as teachers,
provided more mathematics graduates are available.

The West African Examination Council recently proposed Statistics as a subject in its own right in the West African
School Certificate Examination. The syllabus proposed covers
the usual topics found in most elementary courses in Statistics. The new syllabus is due to come into force in the next
year or two. Most of those responding to the questionnaire

had not seen the new syllabus and could not comment on it,
See the Appendix for the syllabus. One respondent felt it was
not wise to introduce Statistics as a whole subject when
it has been difficult to recruit staff to teach even the
present syllabus.

One of the points on which information was sought on the
questionnaire was the suitability of the texts used for teaching Statistics. In the Appendix is given a list of most of the
books used by those responding. The general complaint has
been the unsuitability of most of the books to local conditions.
It was felt that many of the examples were drawn from the experience of pupils in European (particularly the United Kingdom) countries. One person failed to see the relevance of the
birth rate in 19th Century England to pupils in Western Nigeria. Some of the modern texts like the Joint Schools Project
have come in for praise from the point of view of citing local examples. At least one person has mentioned the very well
written book "Living Statistics" by John Bibby and published
by Longmans. It is particularly relevant because it makes use
of local official statistics.

Most of the schools that teach Additional Mathematics start
in the fourth year and have three or four periods a week.

Table 3 below gives some information on the number of candidates entered for the Additional Mathematics paper, the number of candidates entered varied considerably with the school.
In most schools only a handful of pupils were presented for
the examination. The number of passes also varied considerably.

TABLE 3

Statistics of candidates entered for the Additional Mathematics
paper in 1970[3]

Centre	No.of candidates for additional mathematics	No passing	Nor passing with credit	No entered for the whole exam.
1	11	9	6	54
2	10	9	3	62
3	5	5	3	42
4	52	46	34	96
5	19	11	4	49
6	16	15	10	142
7	13	11	9	88
8	8	8	8	58
9	21	20	10	57
10	15	10	4	72

Table 4 below gives the performance of candidates in the centres listed in Table 3 in the three statistics questions in
Paper II of the Additional Mathematics. It is seen that even
in the case of Centre 8 in which all 8 candidates got credit
in Additional Mathematics the performance in Statistics was
not particularly good.

[3]Source: The mark sheets of the West African Examinations
Council, Yaba, Lagos, Nigeria. Used by permission.

TABLE 4

Performance of candidates in some questions in Statistics in the West African School Certificate - 1970

Centre	No.of cand. answering	Question 5 Pictorial representation of data				Question 9 Moving Averages				Question 10 Probability			
		Mean	s.d.	Median	Mode	Mean	s.d.	Median	Mode	Mean	s.d.	Median	Mode
1	9	5.1	4.1	4.0	0.3								
	5					5.4	1.2	6.0	6				
	0									0.0	0.0	0.0	0
2	10	3.7	2.1	3.0	3								
	4					4.8	1.1	5.0	5				
	1									0.0	0.0	0.0	0
3	5	3.0	2.1	4.0	5								
	2					6.0	0.0	6.0	6				
	1									0.0	0.0	0.0	0
4	33	6.7	2.8	7.0	5								
	4					6.0	0.0	6.0	6				
	13									4.7	3.8	6.0	0

Source: The mark sheet of the West African Examination Council, Yaba, Lagos, Nigeria. Used by permission.

5	19	5.3	2.5	6.0	7	5.0	1.2	5.5	6	2.3	5.2	0.0	0
	4											
	7											
6	9	4.4	2.7	5.0	5	6.0	1.0	6.0	5.7	9.2	2.9	10	5,7,10,11,13
	2											
	5											
7	7	6.7	2.4	7.0	1071	6.6	1.7	6.0	6			
	5											
	1											
8	5	3.2	1.8	4.0	1.5	3.0	3.0	3.0	0.6	6.0	0.0	6.0	6
	2											
	1											
9	21	4.4	3.6	4.0	0.4	4.8	2.0	6.0		0.0	0.0	0.0	0
	8											
	0											
10	7	3.5	3.8	3.0	0.3	6.0	0.0	6.0	6	0.0	0.0	0.0	0
	1											
	0											

TABLE 5

Performance of candidates in some questions in Statistics in the West African School Certificate Examination - 1970

Centre	No. of cand. answering	Question 6 Differentiation				Question 7 Quadratic functions				Question 8 Geometry Coordinate			
		Mean	s.d	Median	Mode	Mean	s.d.	Median	Mode	Mean	s.d	Median	Mode
1	4	0.8	1.3	0.0	0	2.9	3.8	2.0	0.3	0.0	0.0	0.0	0
	9												
2	2	0.0	0.0	0.0	0	3.0	1.8	3.0	3	2.0	1.9	2.0	0
	5												
3	9	0.0	0.0	0.0	0	2.3	2.5	1.5	0	2.4	2.2	1.0	1
	7												
4	2	1.9	2.4	1.0	0	3.3	2.8	3.0	3	2.8	1.9	2.0	5
	5												
5	19	1.0	1.4	0.0	0	2.4	3.4	1.0	0	2.1	1.8	2.0	0
	30												
	21												
	10												
	18												
6	14	2.7	3.0	1.5	0	7.6	4.1	10.0	10	7.1	4.2	5.0	5
	6	3.0	1.5	0									
	7												
	9												

Source: The mark sheet of the West African Examination Council, Yaba, Lagos, Nigeria. Used by permission.

Centre	No.of cand.	Question 6 Differentiation				Question 7 Quadratic functions				Question 8 Coordinate geometry			
		Mean	s.d.	Median	Mode	Mean	s.d	Median	Mode	Mean	s.d.	Median	Mode
7	1	1.0	0.0	1.0	1								
	7					8.7	4.2	9.0	5.12				
	8									5.3	3.3	5.0	5
8	2	2.5	2.5	2.5	0.5								
	5					6.0	4.6	6.0					
	4									0.5	0.9	0.0	0
9	11	0.9	1.3	1.0	0								
	11					2.5	2.5	3.0	3				
	11									2.2	1.7	2.0	2
10	5	2.2	2.7	0.0	0								
	6					4.5	1.4	4.5	3.6				
	7									4.4	1.0	5.0	5

Source: The market sheet of the West African Examination Council, Yaba, Logos, Nigeria. Used by permission.

16—Råde

In comparison with table 4 is table 5 which shows the performance of the same set of candidates listed in table 3 in the Pure Mathematics questions of Section B of the Paper II. The sections includes questions in differentation, quadratic functions and elementary coordinate geometry.

In many schools in European countries and in the United States trials are now going on in teaching Statistics to Primary School pupils. This has met with much success in many places. It is quite common to find descriptive statistics in many primary school texts today — The Nuffield project in the United Kingdom and the SMSG project in the United States are examples of this. With the success of these experiments it has been possible to introduce new topics in the statistics and probability curriculum of most secondary schools in these countries. Nigeria has not yet reached this stage. Only recently, however, the National Educational Research Council in Lagos proposed the inclusion of elementary ideas of statistics and probability in the syllabus of primary schools but no serious attempts have so far been made to implement this proposal. It may be that pilot projects will have to be launched by Departments of Education in Nigerian Universities.

A P P E N D I X

1. Traditional Additional Mathematics

Section A

The classification and tabulation of statistical data e.g. population, trade, growth of plants, examination marks. Pictorial representation, e.g.pie charts, bar charts of frequency diagrams, mean, median, mode and quartiles. Measurement of dispersion by range and the interquartile range. Interpretation and application of statistical data.

Section B

Statistics

Histograms including cases of unequal class intervals and its relationship with the cumulative frequency diagram. Moving averages, index numbers, measurement of dispersion by the standard deviation, use of an assumed mean (change of origin) in calculating the standard deviation. The relationship between frequency and probability.

The addition and multiplication laws of probability with simple illustrations.

Frequency and probability distributions. The Binomial Distribution.

2. Modern Mathematics (Alternative "C")

12. Statistics: graphical representation - frequency: mean, mode and median.
13. Simple probability involving equally likely events.

3. Ordinary Level Additional Modern Mathematics

Syllabus	Notes
Statistics Tabulation and graphical representation of data. Mode, median, mean.	Histograms (including unequal class intervals). Including use of assumed mean (change of origin) and calculation of mean from class frequencies.
Cumulative frequency; quartiles and percentiles. Standard deviation. Simple problems on probability, involving both equally and non-equally likely events.	Cumulative frequency curve in simple cases only. The probability of an event considered as the probability of a set.
Sum and product laws. Simple cases of conditional probability.	e.g. drawing balls from box without replacement.

4. Mathematics (Advanced level)

Syllabus	Notes

Statistics
(The questions set will test application of method, use of method and inference rather than mere mechanical calculation.)

29. Frequency distributions, frequency density, histograms, cumulative frequency diagrams.	Pictorial representation.
30. Central tendency; dispersion. Measures of central tendency.	Mean, median, mode, percentiles and quartiles. Range, variance, standard deviation.

Measures of dispersion. Calculation of these measures from a set of numbers and from a frequency distribution.	Calculation of mode from a frequency distribution only.
31. Arrangements and selections. Frequency inter - pretation of probability. Laws of probability, probability density, cu - mulative probability.	Dependent and independent events.
Binomial and Poisson distributions	Situation illustrating these distributions; formal derivations of mean and standard deviation not required.
Rectangular and Normal distributions as examples of continuous distributions.	Application to rounding off errors.
32. Meaning of correlation and regression. Scatter diagrams Graphical treatment of regression. Rank correlation.	Linear regression only

5. Statistics (West African School Certificate).

The Syllabus aims at a treatment of Statistics which shall be broad rather than deep covering the fundamental elementary concepts and including simple calculations.

The topics may be taken in any order found suitable. It seems reasonable to begin with sections A, B, C in that order, but section D, E, F may be taken in any order and it is recommended to us whatever order is favoured.

Problems and illustrations should wherever possible be drawn from the home, school, industry and the community at large. Such problems and illustrations should be meaningful to residents in West Africa.

PAPER I will contain multiple choice questions covering the whole syllabus. Candidates are expected to answer all the questions in this paper.

PAPER II will be divided into two sections, A and B. SECTION A will contain five questions. Candidates are expected to answer all the questions in this section. SECTION B will contain six questions of greater difficulty than these in Section A. Candidates are expected to answer four out of the six questions in Section B.

SYLLABUS

TOPIC	NOTES
A. Collection of Data	
1. Nature of statistical investigations. Range of problems requiring the statistical method.	The inductive nature of sta - tistical reasoning, i.e. find- ing general rules and general characteristics from (a limited number of) specific observations.
2. Sources of statistical data. Objectives. Scope and planning of Census and Surveys; pilot enquiries.	Examples to be drawn from popula- tion, housing, agriculture and industry.
3. Basic ideas of Random and non- random sampling. Quota sampling and other forms of systematic and subjective sampling as ex- amples of non-random samp- ling.	Distinguish between sample and population. Illustrate with simp- le examples.
4. Units and Methods of enumeration.	Person, household as units. Inter- viewer and mail questionnaire as examples of methods of enumeration.
Simple ideas on the design of questionnaires.	Types of questions, e.g. closed, semi-closed and opened. Candidates will be expected to decide, among given alternatives, what type of question is most suitable for a specific enquiry.
	Reference to type of question should be by means of an example and NOT simply by the use of a technical term.
Post-enumeration surveys.	Their aims and merits.
B. Tabulation and Presentation of Data	
5. Frequency distributions. Grouping of data. Relative and percentage frequen - cies.	Use of class intervals (including unequal class intervals).
	Correction for rounded-off data, e.g. a measurement of 6.5.cm means that the reading lies between 6.45 and 6.55cm.

Comulative frequency dis-
tributions

6. Frequency polygons, histograms and cumulative frequency polygons (Ogives).	See also paragraph 8 (Section)

7. Bardiagrams, pie charts, and pictograms. Population maps using dots of varying density.

Graphs of series of observations.	e.g. time series.

C. Measures of Location and Dispersion

8. Arithmetic Mean, Median and mode. Calculation of arithmetic mean for a given set of numbers and for grouped data.	Use and suitability of the different measures of location. Use of the σ (sigma) notation and subscripts. $\sum x = x_1 + x_2 + x_3 + \ldots + x_n$
Finding the median and mode for a set of numbers.	Use of assumed mean and a scaling factor e.g. $y = \dfrac{x - a}{k}$

Estimation of median quartiles and percentiles of grouped data from cumulative frequency graphs.

Estimation of mode from histogram

9. Range, interpercentile ranges, semi-interquartile range; variance and standard deviation for a set of numbers and for grouped data.	Use and suitability of the different measures of dis - persion. Use of assumed mean and scaling factor.

D. Probability and Probability Distributions

10. Meaning of Probability. Numerical Calculation of probabilities in simple situations.	Relative frequency interpretation of probability. Random arrangements and selections.
11. Binomial and Normal distributions. Use of normal distribution tables.	Mathematical derivation of the means and standard deviation not required.
Simple significance testing.	Simple test on binomial populations. Sample inspection.

Confidence intervals for means of normal population.

Test for means of large sample from normal population.

E. Bivariate Data

12. Representation by Scatter diagram. Graphical treatment of regression and correlation. Fitting regression line to scatter diagrams by eye and the determination of its equation $y = mx + c$ by calculating of its slope and reading off the intercept.

Meaning and use of regression coefficient. Mathematical derivation of regression coefficient not required.

13. Definition (by formula) and calcualtion of the product moment correlation coefficient r. Interpretation of r.

Interpretation of positive negative values of r. The significance of values of r close to 1 and values of r near zero.

Rank correlation Spearman's coefficient

$$Q = \frac{1 - 6 \sum d^2}{N(N^2 - 1)}$$

In the case of tied ranks the convention of giving the arithmetic mean rank to each of the equal items will be used.

F. Time Series, Weighted Averages, Index Numbers

14. Elementary analysis of time series. Secular trend and seasonal cycles.

Use of moving averages to remove seasonal variations.

15. Weighted averages

Crude and standardised birth and death rates. Consumer and wholesale price indices (see 16)

16. Index Numbers and their uses. Consumer and wholesale price indices; value, volume, and quantity indices.

6. Statistics in the General Paper (General Studies) – Higher School Certificate Examination

Section B

Interpretation of graphical, diagrammatical or tabular presentation of facts. There will be a choice of questions.

Books

The following is a list of the books used by most of the schools responding:

Loveday, R.: Statistics - A first course, Cambridge University Press.

Loveday, R.: Statistics - A second course, Cambridge University Press.

McIntosh, D. M.: Statistics for the teacher, Pergamom Press

Ferris, V. W. and Busbridge, J. N.: Modern Mathematics for Secondary Schools, Evans Brothers Limited.

Mitchelmore: Joint Schools Project Mathematics, Longmans Green & Co. LTD., London.

Brookes and Dick: An Introduction to Statistical Method, Heinemann, London.

Erricker, B. C.: Elementary Statistics, English Universities Press.Ltd., London.

Gronhmana, B. C.and Carl J.: Principles and Practice of Statistics, George Harrap & Co., Ltd., London.

Walpole, R. E.: Introduction to Statistics, The Macmillan Company of New York.

Pickard, R. E.: Statistics, Cassell., London.

Sherlock, A. J.: Probability and Statistics, Edward Arnold, London.

Mullholland and James: Fundamentals of Statistics, Butterworth., London.

Balmer: Principles of Statistics, Oliver and Boyd,Edinburgh.

Bibby, J.: Living Statistics, Longmans, London.

Tarro Yamane: Statistics An Introductory Analysis.

Moroney, M. J.: Facts from Figures, Penguin Books.

Weatherburn, C. E.: A First Course in Mathematical Statistics, Cambridge University Press.

Spiegel, M. R.: Statistics Theory and Problems, McGraw Hill Book Co., New York.

Harper, W. M.: Statistics, Allen and Unwin.

Rodha, G. W.: Understanding Graphs and Statistics, Nelson, London.

Bolt, R. L.: Examples in Statistics, J. M. Dent & Sons Ltd., London.

Lewis, K. and Ward, H.: Starting Statistics, Longman Group, Ltd., London.

TEACHING OF PROBABILITY AND STATISTICS IN SECONDARY SCHOOLS

IN AUSTRIA

H. LANG

Bundesgymnasium und Bundesrealgymnasium fur Mädchen, Wien

Since the field of mathematical statistics and especially the applications of statistics to different subjects as natural sciences, economics and industry have been growing rapidly during the last 30-40 years, we try to keep pace with this development in Austrian secondary schools. According to the Austrian curriculum for mathematics for secondary schools (excluding technical and commercial schools) we emphasize basic knowledge in probability theory and some elements of mathematical statistics. In this curriculum the following topics are recommended:

5th year (corresponds to grade 9): In connexion with the characteristics of groups, permutations and groups of permutations are discussed.

6th year: Variations and combinations.

7th year: Elements of statistics. Field of application of statistics, histograms, absolute and relative frequencies, classification of data. Mean, mean deviation, variance, standard deviation.

8th year: Introduction to the theory of probability. Outcome space, event, addition theorem, conditional probability, multiplication theorem, random variables.

These topics are considered for those secondary schools, in which mathematics is taught three lessons a week. In "Realistisches Gymnasium" and "Naturwissenschaftliches Realgymnasium" which emphasize natural sciences and have four to five lessons a week, also the binomial and normal distributions are treated including a derivation of the variance of the binomial dis –

With regard to the content of the course, we have to note:

- A frequentist point of view is adopted in teaching what probability is.
- The properties of the probability function on finite fields are derived from the classical definition.
- The notion of random variable is introduced in a heuristical way, in order to get rid of the difficulties involved in the concept of measurability.

The computation of the characteristic values of different finite distributions is developed in connection with classical problems of combinatorics.

- The statistical significance of Chebyshev's inequality is pointed out.
- Special attention is paid to the linear approximation and to the least squares method.

The modification of the curriculum made it necessary that teachers get familiar with these new topics. For this purpose the Romanian Mathematical Society and the Centre of Mathematical Statistics of the Academy of Romania have organized since 1965 summer schools for teachers. An advanced course in probability and statistics for use of teachers has also been published. This course includes detailed proofs, a variety of examples, helpful remarks and also more topics are covered than in the course mentioned above.

By now the opportunity is discussed of teaching the chapters of the so-called descriptive statistics (which do not require proofs) at an earlier level. In this way it would be possible to enlarge the content of the chapter devoted to mathematical statistics by introducing some new topics such as estimation theory and hypothesis testing.

LIST OF THE PARTICIPANTS

Dr. H. Aigner, Sektionsrat, Bundesministerium für Unterricht und Kunst, Sektion für höhere Berufsbildende Schulen, Minoritenplatz 5, Wien, Österreich

Professor B. Benjamin, Director of Statistical Studies, Civil Service College. London, England

Dr. K. G. Brolin, Director, Office of Statistics, Communication Sector, Unesco, Place de Fontenoy, Paris 7e, France

Professor Dr. W. Eberl, Technische Hochschule, Nusswaldgasse 22a 1190 Wien, Österreich

Professor A. Engel, Johann Wolfgang Goethe-Universität, Senckenberganlage 9-11, 6 Frankfurt am Main, Bundesrepublik Deutschland

Mrs. M. Halmos, Mathematical Institute of the Hungarian Academy of Sciences, Reáltanoda u. 13-15, 1053 Budapest, Hungary

Professor P. L. Hennequin, Département de Mathématiques Appliquées, Bolte Postale 45, 63 Aubière, France

Professor W. Kruskal, Chairman, Department of Statistics, University of Chicago, 1118 E. 58th Street, Chicago, Illinois 60637, USA

Professor Dr. Hilde Lang: Untere Donaustraße 23, 1020 Wien, Österreich

Dr. E. Lunenberg, Director, International Statistical Institute, 2 Oostduinlaan, The Hague, Netherlands

Professor F. Mosteller, Department of Statistics, Harvard University, One Oxford Street, Cambridge, Mass. 02138, USA (Chairman of the Conference)

Dr. J. O. Oyelese, Department of Mathematics, University of Ibadan, Ibadan, Nigeria

Dr. L. Peczar, Stadtschulrat für Wien, Dr. Karl Rennerring 1, Wien 1 DR, Österreich

Professor T. Postelnicu, Secretary, Academia Republicii Socialiste România, Colea Victoriei 125, Bucaresti, Romania

Dr. H. Raiffa, International Institute for Applied Systems Analysis, 2361 Laxenburg, Österreich

Dr. L. Råde, Department of Mathematics, Chalmers Institute of Technology, Fack, 40220 Göteborg, Sweden

Dr. C. R. Rao, Secretary and Director, Indian Statistical Institute, 538 Yojana Bhavan, Parliament Street, New Delhi 1, India

Mrs. I. Rucker, Director of Special Programs for the Gifted, State Department of Education, Richmond, Virginia 23216, USA

Professor Dr. L. Schmetterer, Mathematisches Institut der Universität Wien, Strudlhofgasse 4, 1090 Wien, Österreich

Mr. A. Le Thomas, Coutre National d'Information pour le Progrés Economique, 07 La DéFeuse, Paris, France

Mrs. A. Whitfield, International Statistical Institute, 2 Oostduinlaan, The Hague, Netherlands (secretary of the conference)